SECOND-YEAR TECHNICIAN
MATHEMATICS

Other books in this series
First-year Technician Mathematics: Rhys Lewis
Third-year Technician Mathematics and Applications: Rhys Lewis

Books by the same author
Electronic Systems for Radio, Television, and Electronics
Mechanics

Books of related interest
Linear Electronic Circuits and Systems: *G.D. Bishop*
Digital Electronic Circuits and Systems: *Noel M. Morris*
Essential Formulae for Electronic and Electrical Engineers: *Noel M. Morris*
Electrical Circuits and Systems: *Noel M. Morris*
The Electrical Principles of Telecommunications: *R. Lowe and D. Nave*
Basic Electrotechnology: *H. Cotton*

SECOND-YEAR TECHNICIAN MATHEMATICS

for Electrical, Electronics and Telecommunications Students

RHYS LEWIS
B.Sc. Tech., C.Eng., M.I.E.E.
Senior Lecturer, Openshaw Technical College, Manchester

First published 1975 by
THE MACMILLAN PRESS LTD
London and Basingstoke
Associated companies in New York Dublin
Melbourne Johannesburg and Madras

SBN 333 17846 7

printed and bound in Great Britain by
Redwood Burn Limited *Trowbridge and Esher*

Contents

Preface

This book is the second in a series of three books covering the mathematical syllabuses of courses leading to Electrical and Electronic Technician status and ultimately to Technician Engineer status. At the present time these courses include the City and Guilds of London Institute courses 280, 281, 285, 270 and 272. The titles of the books, breaking the total syllabuses into four years, stem from the organisation of these courses. With the advent of the Technician Education Council, and subsequent reorganisation of courses, it may be found that the breakdown by year may change; however, it is anticipated that the total content of all three volumes will more than adequately cover the total syllabus content of all electrical and electronic technician courses both present and proposed. All three books contain a large number of worked examples and exercises for the students to complete and end with a number of typical examination papers covering the content of each volume.

The assistance and advice of Mr Noel M. Morris, the series editor, is gratefully acknowledged. In addition, I would particularly like to thank Mrs L. J. E. Jones for typing the manuscript and, as always, my wife for her help and encouragement.

Openshaw, Manchester   RHYS LEWIS

1 Arithmetic and Aids to Calculation

SYSTEMS OF NUMBERS

The principles common to most systems of numbers, which were considered in detail in *First-year Technician Mathematics*, are presented again below.

Most systems of numbers have a *base* or *radix* and a total count is represented by a series of digits, each of which gives the number of times a power of the base is contained in the total. The particular power is governed by the position of the digit in the overall number. The number of symbols or digits in the system is equal to the radix. Thus in the decimal or denary system (base ten) there are ten digits (0 to 9) and the decimal number 1579, for example, means

$$9 \times \text{ten raised to the power 0}$$

$$7 \times \text{ten raised to the power 1}$$

$$5 \times \text{ten raised to the power 2}$$

$$1 \times \text{ten raised to the power 3}$$

that is

$$1 \times (\text{ten})^3 + 5 \times (\text{ten})^2 + 7 \times (\text{ten})^1 + 9 \times (\text{ten})^0$$

and since any number raised to the power 0 is unity this equals

$$1000 + 500 + 70 + 9$$

The base itself is always represented by the digits 10 since this means $1 \times (\text{base})^1$ plus $0 \times (\text{base})^0$, that is, equals the base.

Similarly the octal system (base eight) has eight digits (0 to 7) and the number 765, for example, means

$$7 \times (\text{eight})^2 + 6 \times (\text{eight})^1 + 5 \times (\text{eight})^0$$

that is, *in decimal numbers*

$$(7 \times 64) + (6 \times 8) + 5 \text{ or } 501$$

Again the base eight is written as 10 in the octal system.

The binary system

The binary system has only two digits, 1 and 0, and since these are easily represented in electrical or electronic circuits by the conditions 'on' and 'off', this system is widely used in computing and other logic systems. The base or radix of the binary system is two, written as 10 in the binary system. Binary numbers for decimal numbers from 1 to 16 are presented below.

Decimal	Binary
0	0
1	1
2	10
3	11
4	100
5	101
6	110
7	111
8	1000
9	1001
10	1010
11	1011
12	1100
13	1101
14	1110
15	1111
16	10000

Taking decimal 9 as an example, 1001 means

$$1 \times (\text{two})^3 + 0 \times (\text{two})^2 + 0 \times (\text{two})^1 + 1 \times (\text{two})^0$$

that is, in decimal

$$8 + 0 + 0 + 1$$

Conversion from binary to decimal is easily effected by adding the powers of two as indicated by whether the digit is 1 or 0 and by the position of the digit in the overall number. For example, binary 1011101101 means, in decimal

$$(1 \times 2^9) + (0 \times 2^8) + (1 \times 2^7) + (1 \times 2^6) + (1 \times 2^5) + (0 \times 2^4)$$
$$+ (1 \times 2^3) + (1 \times 2^2) + (0 \times 2^1) + (1 \times 2^0)$$

that is

$$512 + 0 + 128 + 64 + 32 + 0 + 8 + 4 + 0 + 1 \text{ or } 749$$

Conversion from decimal to binary is achieved by repeated division by 2 and noting the remainder. For example, to convert 749 to binary

Remainder

```
2 ) 749
      374        1
      187        0
       93        1
       46        1
       23        0
       11        1
        5        1
        2        1
        1        0
START
```

Start at the final 1 in the divide column then read up to give 1011101101 as in the previous example.

Basic arithmetical processes in the binary system

Addition

There are four basic rules, which are

$$0 + 0 = 0$$
$$0 + 1 = 1$$
$$1 + 1 = 0 \text{ carry } 1$$
$$1 + 1 + 1 = 1 \text{ carry } 1$$

The 'carry 1' in the last two lines means that 1 is carried over to the next column on the left when the numbers are arranged one beneath the other. Binary numbers should only be added in pairs of numbers at any one time, otherwise the 'carry' requirement may become confusing, that is, it may be necessary to carry 1 to a column further away than one place to the left.

Example 1.1

Add the binary numbers 101101, 111010, 1011 and check the answer by decimal conversion of all numbers.

First, add 101101 to 111010 as follows: write one number beneath the other and commence adding at the last column on the right side.

Column 7	Column 6	Column 5	Column 4	Column 3	Column 2	Column 1
	1	0	1	1	0	1
	1	1	1	0	1	0
1	1	0	0	1	1	1

For columns 1, 2 and 3 we have $1 + 0$ or $0 + 1$ giving a digit 1 in each column. Column 4 is $1 + 1$ giving 0 in the answer, carry 1 to column 5. Column 5 becomes $0 + 1 + 1$ (carried) giving 0 in column 5 answer, carry 1 to column 6. Column 6 becomes $1 + 1 + 1$ (carried) giving 1 in column 6 answer, carry 1 to column 7 answer. (The above layout is for convenience of explanation and need not be followed once the basic principles are understood.) Now add the answer 1100111 to the remaining number 1011 using the routine outlined below.

$$
\begin{array}{r}
1100111 \\
1011 \\
\hline
1110010
\end{array}
$$

Column 1	$1 + 1$	$= 0$	carry 1 to column 2
Column 2	$1 + 1 + 1$ (carried)	$= 1$	carry 1 to column 3
Column 3	$1 + 0 + 1$ (carried)	$= 0$	carry 1 to column 4
Column 4	$0 + 1 + 1$ (carried)	$= 0$	carry 1 to column 5
Column 5	$0 + 1$ (carried)	$= 1$	
Column 6	$1 + 0$ (understood)	$= 1$	
Column 7	$1 + 0$ (understood)	$= 1$	

The final answer is thus 1110010, which may be checked by conversion as follows.

$$101101 \text{ in decimal means } 32 + 8 + 4 + 1 \quad \text{that is, } 45$$
$$111010 \text{ in decimal means } 32 + 16 + 8 + 2 \quad \text{that is, } 58$$
$$1011 \text{ in decimal means } 8 + 2 + 1 \quad \text{that is, } 11$$

The answer 1110010 in decimal means $64 + 32 + 16 + 2$, that is, 114 and since $45 + 58 + 11 = 114$ the answer is correct.

Subtraction

A set of basic rules for binary subtraction may be derived in a similar form to those derived for binary addition. However, since these involve 'borrowing' 1 from another column whenever 1 is subtracted from 0 this subtraction process is often found confusing. A better method is to use the addition process 'in reverse' by determining what number must be added to the number that is to be subtracted in order to equal the number from which that subtraction is to take place. To clarify this consider, for example, subtracting 1111 from 11001; the process reduces to finding the number which must be added to 1111 to equal 11001.

Example 1.2

Subtract 1111 from 11001.

	Column 5	Column 4	Column 3	Column 2	Column 1
Row 1	1	1	0	0	1
Row 2		1	1	1	1
Row 3	0	1	0	1	0

In column 1 consider what number must be added to the 1 in the number being subtracted (row 2) to equal the 1 in the number from which the subtraction is to take place (row 1). The answer is 0 which is placed in column 1, row 3.

Next consider column 2. What number must be added to the 1 in row 2 to equal the 0 in row 1? Since 1 + 1 = 0 carry 1, the answer is 1, which is put in column 2, row 3.

In column 3 row 2 we now have 1 plus 1 carried from column 2. What number must be added to this 1 + 1 to give 0 in column 3, row 1? The answer is 0 since 1 + 1 = 0 carry 1 (the 1 to be carried to column 4). Write 0 in column 3, row 3.

In column 4, row 2 we now have 1 + 1. What number must be added to the 1 + 1 in row 2 to give the 1 in row 1? Clearly this must be 1 since 1 + 1 + 1 = 1 carry 1 (the 1 to be carried to column 5). Write 1 in row 3, column 4.

Finally, in column 5, row 2 we now have a 1 carried from column 4. What number must be added to 1 to give the 1 in row 1? The answer is 0. Write 0 in row 3, column 5.

The overall answer is thus 01010 or 1010 since a 0 is the first digit and may be omitted. Check by adding together rows 2 and 3.

	Row 2		Row 3	Carried		Row 1	Carry	
Column 1	1	+	0		=	1		
Column 2	1	+	1		=	0	1	to column 3
Column 3	1	+	0	+ 1	=	0	1	to column 4
Column 4	1	+	1	+ 1	=	1	1	to column 5
Column 5			0	+ 1	=	1		(complete)

The student is advised to read through this example again and ensure each line is fully understood. More examples with reducing detailed explanation follow.

Example 1.3

Subtract 11011 from 111010.

$$
\begin{array}{r}
111010 \\
\underline{11011} \\
\underline{11111}
\end{array}
$$

Column 1 (right hand side)

> To give 0 in row 1, 1 must be added to the 1 in row 2.
> Write 1 in row 3. This will carry 1 to column 2 when rows 2 and 3 are added.

Column 2   To give 1 in row 1, 1 must be added to the 1 in row 2 and the 1 from column 1. Write 1 in row 3. This will carry 1 to column 3 when rows 2 and 3 are added.

Column 3   To give 0 in row 1, 1 must be added to the 1 in row 2 carried from column 2. Write 1 in row 3. This will carry 1 to column 4 when rows 2 and 3 are added.

Column 4   To give 1 in row 1, 1 must be added to the 1 in row 2 and the 1 carried from column 3. Write 1 in row 3. This will carry 1 to column 5 when rows 2 and 3 are added.

Column 5   As column 4. Write 1 in row 3. This will carry 1 to column 6. Row 1 column 6 is 1. No further digits needed.

Subtracting in denary gives $58 - 27$ which is 31. Binary equivalent of 31 is 11111, thus the answer is correct.

Example 1.4

Subtract 110110 from 111000.

$$
\begin{array}{r}
111000 \\
\underline{110110} \\
\underline{000010}
\end{array}
$$

Example 1.5

Subtract 111011 from 1110000.

$$
\begin{array}{r}
1110000 \\
\underline{111011} \\
\underline{110101}
\end{array}
$$

Multiplication

The rules for multiplication are

$$0 \times 0 = 0$$
$$0 \times 1 = 0$$
$$1 \times 0 = 0$$
$$1 \times 1 = 1$$

A calculation involving multiplication is written in much the same way as with any other number system, the number in the top line being first multiplied by the last digit of the number in the next line, then by the next to the last digit and so on, successive sub-totals being moved one place to the left as the multiplication proceeds. The sub-totals are then added to give the answer (taking two lines at a time to avoid confusion).

Example 1.6

Evaluate 1011×1110.

First number, row 1	1011	
Second number, row 2	1110	
First number × last digit row 2 (0)	0000	row three
First number × next digit left row 2 (1)	10110	row four
First number × next digit left row 2 (1)	101100	row five
First number × next digit left row 2 (1)	1011000	row six
	Add 10011010	

Note that as the first number moves progressively one place to the left in rows 4, 5 and 6 the remaining spaces in these rows are filled with 0s. This assists 'lining up' of columns and helps avoid error when adding.

Sometimes when adding three or more lines gives more than three 1s it is advisable to add two rows at a time to avoid 1 being carried more than one place to the left. ($1 + 1 + 1 = 1$ carry 1 but $1 + 1 + 1 + 1 = 0$ carry 0 to the next column left and carry 1 to the next column again to the left. This can cause error.) See examples 1.7 and 1.8.

Example 1.7

Multiply 10111 by 11011.

	10111	
	11011	
Row 3	10111	
Row 4	101110	
Row 5	0000000	*Note* last digits are 0 in these
Row 6	10111000	rows as 10111 moves to the left
Row 7	101110000	

ADD	Row 3	10111
	Row 4	101110
		1000101

ADD	Row 6	10111000
	Row 7	101110000
		1000101000

ADD		1000101
		1000101000
		1001101101

A denary check gives 10111 × 11011 is 23 × 27 which equals 621; the binary equivalent of 621 is 1001101101, thus the answer is correct.

Division

The rules are

$$0 \div 1 = 0$$
$$1 \div 1 = 1$$

The layout of a calculation involving binary division is much the same as for other systems.

Example 1.8

Divide 100100 by 110.

```
                110
        110/100100
               110
               110
               110
               000
```

Notice that the procedure is similar to 'long division' with numbers from other systems. 110 will not divide into 1, 10, 100 and so the first set of numbers into which 110 may be divided is 1001 into which 110 is divided once. Write 1 above the line over the right-hand digit of 1001 and subtract 110 from 1001 to give 11. Bring down the next digit 0 to make 110 divides once. Write 1 above the line and subtract 110 from 110. There is no remainder, but since there is a remaining 0 in 100100, write 0 above the line to give an answer 110. A denary check gives $36 \div 6 = 6$ which is, of course, correct.

Example 1.9

Divide 100111 by 11.

$$\begin{array}{r} 1101 \\ \hline 11/100111 \\ \underline{11} \\ 11 \\ \underline{11} \\ \overline{011} \\ \underline{11} \\ \overline{00} \end{array}$$

In this example 11 will not divide into 1 or 10 but divides into 100 once. Write 1 in the answer, subtract 11 from 100 to give 1. Bring 1 down to make 11. 11 divides into 11 once, write 1 in the answer. There is no remainder but figures other than 0 still remain in 100111. Bring down 1 to make 01. 11 will not divide into 01 so write 0 in the answer. Bring down the last 1 to make 011 (that is, 11) and divide by 11 to give 1 in the answer, a remainder of zero, and no figures left in 100111. A denary check gives the division as 39 by 3 giving 13, which is the denary equivalent of 1101; thus the answer is correct.

Non-integral numbers: binary fractions

As explained in detail in *First-year Technician Mathematics,* numbers other than whole numbers (integers) may be expressed in binary form (as indeed they may be in any numbering system) by use of a radix point called the 'decimal point' in the denary system and the 'binary point' in the binary system.

Digits before the point indicate the number of times the radix raised to a positive power is contained in the total count, and similarly digits after the point indicate the number of times the radix raised to a negative power is contained in the total count. Thus, for the binary system, 110.101 means

$$(1 \times 2^2) + (1 \times 2^1) + (0 \times 2^0) + (1 \times 2^{-1}) + (0 \times 2^{-2}) + (1 \times 2^{-3})$$

using the symbol 2 to indicate two for convenience (as explained earlier the symbol 2 is not used in the binary system, being replaced by 10, the symbol for the radix in any system). Thus in denary, 110.101 means,

$$4 + 2 + 0 + \frac{1}{2} + 0 + \frac{1}{8}$$

or

$$4 + 2 + 0.5 + 0.125 \text{ that is, } 6.625$$

To convert from decimal to binary for such numbers a convenient method is to repeatedly double the denary fraction, and to write 1 in the binary equivalent if the result of this doubling produces a 1 *before* the decimal point in the decimal number.

Example 1.10

Convert 0.65625 to binary.

Double 0.65625 to give 1.3125; there is a 1 before the point so write 1 in the binary fraction giving 0.1. Double 0.3125 to give 0.625; no 1 before the point so write 0 in the binary fraction to give 0.10. Double 0.625 to give 1.25; there is a 1 before the point so write 1 in the binary fraction to give 0.101. Double 0.25 to give 0.5; write 0 to give 0.1010. Double 0.5 to give 1.0; write 1 to give 0.10101. The binary equivalent of 0.65625 is therefore 0.10101 which means (1×2^{-1}) + $(1 \times 2^{-3}) + (1 \times 2^{-5})$, that is $0.5 + 0.125 + 0.03125$.

Example 1.11

Convert 0.34375 to binary.

	Binary	
$2 \times 0.34375 = 0.6875$	0.0	
$2 \times 0.6875 = 1.375$	0.01	
$2 \times 0.375 = 0.75$	0.010	
$2 \times 0.75 = 1.5$	0.0101	
$2 \times 0.5 = 1.0$	0.01011	*Answer*

Where the decimal fraction is not exactly equal to the sum of reducing powers of two, as in the preceding examples, the process proceeds until the required number of places is reached. The rules of addition, subtraction, division and multiplication apply equally to non-integral numbers as shown below.

Example 1.12

Convert 1.437 and 0.652 to binary and then evaluate the following: (a) 1.437 + 0.652, (b) 1.437 − 0.652, (c) 1.437 × 0.652, (d) 1.437/0.652.

Conversion	Binary		Binary
$2 \times 0.437 = 0.874$	0.0	$2 \times 0.652 = 1.304$	0.1
$2 \times 0.874 = 1.748$	0.01	$2 \times 0.304 = 0.608$	0.10
$2 \times 0.748 = 1.496$	0.011	$2 \times 0.608 = 1.216$	0.101
$2 \times 0.496 = 0.992$		$2 \times 0.216 = 0.432$	0.1010
$= 1$ (nearest)	0.0111	$2 \times 0.432 = 0.864$	0.10100

Take 0.0111 to represent 0.437, so 1.437 is 1.0111 in binary. Take 0.101 to represent 0.652.

(a) Addition	Denary		Binary	
	1.437		1.0111	
	0.652		0.1010	
	2.089	*Answer*	10.0001	*Answer*

(b) Subtraction

	Denary		Binary	
	1.437		1.0111	
	0.652		0.1010	
	0.785	*Answer*	0.1101	*Answer*

(c) Multiplication

	Denary		Binary	
	1.437		1.0111	
	0.652		0.1010	
	2874		101110	
	7185		10111000	
	8622		0.11100110	*Answer*
	0.936924	*Answer*		

To determine the position of the radix point count the total number of figures after the point in each number when arranged as shown (4 in one + 4 in the other gives 8) then beginning at the right-hand side of the answer 11100110, move this number of places to give 0.1110011. Similarly with the decimal calculation move 6 places along 936924 to give 0.936924.

(d) Division

First multiply both numbers by a sufficient power of the radix until the divisor becomes a whole number, that is

$$\frac{1.437}{0.652} = \frac{1437}{652}$$

and

$$\frac{1.0111}{0.101} = \frac{1011.1}{101}$$

Now divide as normal.

Denary	Binary
2.2039	10.01001
652/1437	101/1011.1
1304	101
1330	0110
1304	101
2600	1000
1956	101
6440	110
5868	etc.
572	
etc.	
Answer 2.2039	*Answer* 10.01001

When the binary results are checked against the decimal results it will be apparent that an error exists. This is due to the approximation of taking 1.0111 as 1.437 and 0.101 as 0.652. In fact 1.0111 is $1 + 0.25 + 0.125 + 0.0625$, that is, 1.4375; and 0.101 is $0.5 + 0.125$, that is, 0.625. It can be seen that a greater number of places is required in binary calculation to obtain the same degree of accuracy as can be obtained with a denary calculation.

Exercise 1.1

(1) Convert to binary (a) 27   (b) 453   (c) 1762   (d) 18 789.

(2) Convert to decimal (denary)   (a) 0111011   (b) 1101011
(c) 1011 1011 1011.

(3) Evaluate (a) $(101110 + 1011 + 111011)$ (b) $111011 + 1001 - 110$
(c) 10111×1101 (d) $11011 \div 11$. Check by conversion to denary.

(4) Evaluate (a) $(10111 \times 11) + (10111 - 1101)$
(b) $(110111 + 101 - 10111) \times 111$ (c) $(11101 - 11) \div 1001$

(5) Convert to binary (a) 0.734   (b) 5.678   (c) 13.214   (take 3 figures after the radix point).

(6) Convert to decimal (a) 0.11101   (b) 1110.110   (c) 10110.11011.

(7) Evaluate (a) $110.101 + 110.10$ (b) $110110.1011 - 110.1010$.

(8) Evaluate (a) 110.11×10.11 (b) $11011.101 \div 11.01$.

(9) Evaluate $\dfrac{101.11 \ \times \ 110.01}{11101.11 \times \ \ 0.11}$

(10) Evaluate $(110.11 - 100.111) \times 0.11 + (1101.1 + 0.111) \div 0.111$.

USE OF LOGARITHMS

The basic theory of logarithms was presented in *First-year Technician Mathematics*. In this book we are concerned with further applications, particularly of logarithms to a base other than ten. First, however, the basic theory will be reviewed.

Review of basic theory of logarithms

The logarithm of any number to a particular base is the power to which the base must be raised to equal the number. Using symbols, if

$$\log_b x = y$$

then

$$b^y = x$$

Expressing numbers in logarithmic form we may, from the laws of indices, use the simpler processes of addition and subtraction of logarithms to carry out multiplication and division of the original numbers. For example, the laws of indices state that if a base b is raised to powers y_1 and y_2 then

$$b^{y_1} \times b^{y_2} = b^{y_1 + y_2}$$
$$b^{y_1} \div b^{y_2} = b^{y_1 - y_2}$$

and we see that the powers y_1 and y_2 are added when the numbers b^{y_1} and b^{y_2} are multiplied and y_2 is subtracted from y_1 when the number b^{y_1} is divided by the number b^{y_2} to give the resultant power to which b must be raised to give the required answer.

To carry out multiplication or division, or other mathematical operation other than addition or subtraction, the logarithms of the numbers are first determined from tables and at the conclusion of the calculation the answer is reconverted from logarithm to number, using tables of antilogarithms. The most commonly used base is ten and logarithms to the base ten are called common logarithms. Common logarithms have certain inherent advantages which are reviewed below. Other bases are equally possible, and one particularly useful one is the constant e (having the value 2.7183 to four places). Such logarithms, discussed later, are called natural, hyperbolic or Naperian logarithms (after the mathematician John Napier).

Use of common logarithms

All logarithms are made up of two parts, the characteristic which lies before the decimal point and which may be positive or negative, and the mantissa, which lies after the decimal point and is always positive. When a characteristic is negative the minus sign is written over the number which is then referred to as 'bar', for example the characteristic -2 is written $\bar{2}$ and called 'bar 2'. With common logarithms the mantissa of any number that has the same digits in the same order is always the same, only the characteristic is different and this is determined by the position of the decimal point. Thus it is only necessary to have one set of tables of logarithms.

The mantissa of the logarithm of any number is obtained by locating the digits of the number using the appropriate columns and row, and difference columns where necessary, the position of the decimal point being ignored until it is time to determine the logarithm characteristic. For logarithms of numbers greater than one the characteristic is positive and equals one less than the number of digits *preceding* the decimal point. For logarithms of numbers less than one the characteristic is negative and equals one more than the number of noughts between the decimal point and the first digit greater than zero *after* the decimal point.

Skill in the handling of common logarithms is fairly easily acquired with practice, but particular care must be taken when handling bar numbers (especially when subtracting) and when logarithms that are a mixture of positive and negative numbers are converted to wholly negative numbers in problems that involve raising to non-integral powers. Some more difficult examples are given below and problems requiring use of these techniques are set in the exercise that follows.

Example 1.13

Divide 0.00314 by 0.0731 using common logarithms

The mantissa of log 0.00314 is 4969 by locating 31 (first column) and following the row until the column headed by 4 is reached. Since 0.00314 is less than one the logarithm has a negative characteristic. Two noughts follow the decimal point before the first digit greater than zero (3), thus the characteristic is bar 3 written $\bar{3}$ and log 0.00314 is $\bar{3}$.4969. Similarly log 0.0731 is $\bar{2}$.8639.

Division of 0.00314 by 0.0731 requires subtraction of $\bar{2}$.8639 from $\bar{3}$.4969. Subtraction proceeds as normal but care must be taken when the characteristics are reached; bar 2 becomes + 2 since the logarithm $\bar{2}$.8639 is being subtracted and when one is 'borrowed' frcm the column to the left (when 8 in $\bar{2}$.8639 is subtracted from the 4 in $\bar{3}$.4969) this becomes negative if replaced in either the lower row or upper row*. Thus we have either

$$(\bar{3}) - (\bar{2} + 1) \qquad \text{that is, } -3 + 2 - 1$$

or

$$(\bar{3} - 1) - (\bar{2}) \qquad \text{that is, } -3 - 1 + 2$$

which both give − 2, that is, bar 2, as the characteristic of the logarithm of the answer.

*Depending on how the student has learned basic subtraction; the processes and implications are fully explained in *First-year Technician Mathematics*.

The solution is laid out

Number	Log
0.00314	$\bar{3}.4969$
0.0731	$\bar{2}.8639$
0.04295	$\bar{2}.6330$

Example 1.14

Use logarithms to simplify

$$\frac{0.00721 \times 0.0356}{0.0005 \times 0.0041}$$

Number	Log	
0.00721	$\bar{3}.8579$	} Add
0.0356	$\bar{2}.5514$	
	$\bar{4}.4093$	
0.0005	$\bar{4}.6990$	} Add
0.0041	$\bar{3}.6128$	
	$\bar{6}.3118$	
	$\bar{4}.4093$	} Subtract
	$\bar{6}.3118$	
Answer	2.0975	
125.1		

The subtraction of the lower characteristic from the upper gives $\bar{4} - \bar{6}$, that is, $-4 - (-6)$ which equals $-4 + 6$ or 2. *Note* When using tables of antilogarithms the reverse process is used in which only the mantissa is located in tables, the characteristic being used to determine the position of the decimal point.

Example 1.15

Evaluate $0.0047^{0.678}$.

To solve this problem using logarithms, the logarithm of 0.0047 is first determined from tables then multiplied by 0.678 (using logarithms again if required) and the answer obtained by antilogging the result. Great care and also an understanding of the basic nature of a logarithm are needed (in this case, negative characteristic, positive mantissa) and if logarithms are used twice, the second time for multiplication by 0.678, antilogging must also be carried out twice. The detailed solution follows.

Number	Log
0.0047	$\bar{3}.6721$

Now multiply $\overline{3}.6721$ by 0.678, that is

$$(-3 + 0.6721) \times 0.678$$

or

$$-2.3279 \times 0.678$$

Number	Log	
2.3279	0.3670	Add
0.678	$\overline{1}.8312$	
1.579	0.1982	

which equals -1.579 (ignore minus sign when taking logs) that is, $-2 + 0.421$, rewriting with negative characteristic, $\overline{2}.421$.
Thus

$$\text{antilog } \overline{2}.421 = 0.0047^{0.678}$$
$$= 0.02636$$

Note When reconverting -1.579 (wholly negative) to $\overline{2}.421$ (partly negative, partly positive) increase the number preceding the decimal point in -1.579 by one to obtain the characteristic (a bar number) and subtract the part of -1.579 after the decimal point (that is, 0.579) from 1 to obtain the mantissa.

Exercise 1.2
Use common logarithms to evaluate the following

(1) $\dfrac{0.0034 \times 0.017}{0.0562}$ (2) $\left(\dfrac{0.004}{0.078}\right)^{1/2}$ (3) $\dfrac{(0.0067 \times 0.021)^{1/2}}{1.742}$

(4) $0.0789^{0.46}$ (5) $0.7821^{-0.745}$ (6) $(0.00789 \times 0.0145)^{-0.2}$

(7) $4(0.0175 \div 0.345)^{0.7}$ (8) $0.734^{0.2} \times 0.2^{0.734}$

(9) $\left(\dfrac{0.0075}{1532}\right)^{-0.072}$

Natural logarithms

As previously stated, and as explained in detail in *First-year Technician Mathematics*, logarithms to the base ten have the distinct advantage of having the same mantissa for all numbers that have the same digits in the same order. Only the characteristic changes, its value being determined by the position of the decimal point.

Thus, for example

$$\log 0.005431 = \bar{3}.7349$$
$$\log 0.05431 \ \ = \bar{2}.7349$$
$$\log 0.5431 \ \ \ = \bar{1}.7349$$
$$\log 5.431 \ \ \ \ = 0.7349$$
$$\log 54.31 \ \ \ \ = 1.7349$$
$$\log 543.1 \ \ \ \ = 2.7349$$

and so on. The reason for this is that any number may be expressed as the product of a number lying between 1 and 9.999 and a multiplier which is a power of ten. Since the logarithms are to the base ten, the logarithm of that multiplier is the power itself and this in turn becomes the characteristic. The mantissa is always the same for the same order of digits and is given in the tables, only one set of which is needed. Using the above example

$$\log 0.005431 = \log (5.431 \times 10^{-3})$$
$$= \log 5.431 + \log 10^{-3}$$
$$= 0.7349 + (-3) \text{ written as } \bar{3}.7349$$

and

$$\log 54.31 \ \ \ = \log (5.431 \times 10)$$
$$= \log 5.431 + \log 10$$
$$= 0.7349 + 1$$
$$= 1.7349$$

This advantage of common logarithms is due to the fact that the base of the logarithm is the same as the base of our number system, that is, ten. Natural logarithms are to the base e whereas the number system in which they are written is to the base ten and so the mantissa and characteristic will change every time the decimal point is moved.

Consider the logarithm of 0.005431 to the base e (written ln 0.005431 or $\log_e$ 0.005431)

$$\ln 0.005431 = \ln (5.431 \times 10^{-3})$$
$$= \ln 5.431 + \ln (10^{-3})$$

and since $\ln (10^{-3})$ is not $- 3$ but is itself composed of a part before and after the decimal point, ln 5.431 will need to be changed before and after the decimal point to equal ln 0.005431. Similarly since

$$\ln 0.05431 = \ln (5.431 \times 10^{-2})$$
$$= \ln 5.431 + \ln (10^{-2})$$

and ln (10^{-2}) is not -2 but is a number composed of two parts, these will have to be added to ln 5.431 to give ln 0.05431. Thus ln 0.005431 and ln 0.05431 do *not* have the same mantissa and the simple rules for obtaining common logarithms (obtaining the mantissa from a single set of tables and determining the characteristic from the position of the decimal point) have to be amended when natural logarithms are required.

The process of finding natural logarithms has in fact already been outlined above. The number is expressed in standard form (one figure before the decimal point) with a suitable multiplier consisting of ten raised to the necessary power. The natural logarithm of the standard-form number obtained from one set of tables is then added to the natural logarithm of the multiplier, obtained from a second set of tables (usually given on the same page as the first in most books of tables), to give the final result.

Example 1.16

Find the natural logarithm of the following (a) 0.005431   (b) 0.05431   (c) 0.543₁
(d) 5.431   (e) 54.31   (f) 543.1.

Natural logarithms are usually given for numbers from 1 to 10 using rows, columns and difference columns read in the same way as for common logarithms, together with a separate set of values of natural logarithms for powers of ten. This set gives one row showing the value of n and a second row showing ln 10^n or ln 10^{-n}. The number shown beneath a particular value of n is the natural logarithm of 10^n or 10^{-n} as indicated by the first column, second row.

(a) $$\ln 0.005431 = \ln (5.431 \times 10^{-3})$$
$$= \ln 5.431 + \ln (10^{-3})$$
$$= 1.6921 + \overline{7}.0922 \quad \text{(from tables)}$$
$$= \overline{6}.7843$$

The student may recall that bar numbers, that is, numbers having a negative characteristic but positive mantissa, are used with common logarithms to maintain the advantage of having the same mantissa for all numbers that have the same figures in the same order regardless of the position of the decimal point. One may therefore wonder why bar numbers are used for natural logarithms where this advantage is not available. The reason is that if they are not used the logarithm becomes wholly negative and the ease of handling logarithms, which one acquires with practice, may become impaired because a different presentation of logarithm is used for different bases. This can be particularly troublesome in a problem involving the use of both natural and common logarithms. Consequently the bar-number notation is retained for natural logarithms and ln 0.005431 is written as $\overline{6}.7843$ instead of $-6 + 0.7843$, that is, -5.2157. The remainder of this example is answered without comment but the reader is advised to check the figures by using tables.

(b) $\qquad$ ln 0.05431 = ln (5.431×10^{-2})
$$= \ln 5.431 + \ln (10^{-2})$$
$$= 1.6921 + \bar{5}.3948$$
$$= \bar{3}.0869$$

(c) $\qquad$ ln 0.5431 = ln (5.431×10^{-1})
$$= \ln 5.431 + \ln (10^{-1})$$
$$= 1.6921 + \bar{3}.6974$$
$$= \bar{1}.3895$$

(d) $\qquad$ ln 5.431 = 1.6921

(e) $\qquad$ ln 54.31 = ln (5.431×10)
$$= \ln 5.431 + \ln 10$$
$$= 1.6921 + 2.3026$$
$$= 3.9947$$

(f) $\qquad$ ln 543.1 = ln (5.431×10^{2})
$$= \ln 5.431 + \ln (10^{2})$$
$$= 1.6921 + 4.6052$$
$$= 6.2973$$

The example shows clearly that although all the given numbers (0.005431, 0.05431, etc.) have the same figures in the same order, the mantissa of the natural logarithms differs in all cases, which it would *not* do with common logarithms.

Antilogging of natural logarithms

Antilogging of natural logarithms is not as straightforward as antilogging of common logarithms since the method used to find the logarithm must be reversed and two sets of tables are involved. The natural logarithm must first be adjusted by subtracting the natural logarithm of an appropriate power of ten so that the adjusted value lies within the tables of natural logarithms of numbers from 1 to 10. These tables are then read 'in reverse'.

Example 1.17

Antilog the following natural logarithms

(a) 1.6732 $\qquad$ (b) 7.8914 $\qquad$ (c) $\bar{6}.5132$

(a) This logarithm lies in the main tables. The nearest logarithm *below* 1.6732 is 1.6715 for which the number is 5.32 (obtained by reading the first column in the row, to give 5.3, and the head of the column in which 1.6715 lies, to give 5.32). The difference to be added to the last two figures of 1.6715 to give 1.6732 is 17, and this is found in the last difference column along the row in which 1.6715 lies. The last difference column is headed by 9 and so the number is 5.329. Such an exercise *must* always be checked by finding the logarithm of the number, that is, by reading the tables 'forward' in the usual manner.

(b) 7.8914 does not lie in the main tables since the logarithms contained there start at 0 (ln 1) and finish at 2.3026 (ln 10). Examine the small table giving natural logarithms of 10^n. One of these must be subtracted from 7.8914 to give a logarithm which does lie in the main tables.

$$\text{For } n = 2, \ln (10^2) = 4.6052$$
$$n = 3, \ln (10^3) = 6.9078$$
$$n = 4, \ln (10^4) = 9.2103$$

Subtracting 4.6052 from 7.8914 gives 3.2862 which lies outside the range of the tables (0 to 2.3026). Similarly, subtracting 9.2103 from 7.8914 gives −1.3189 which also does not lie in the tables. However, subtracting ln 10^3, that is, 6.9078 from 7.8914 gives 0.9836 which does lie in the tables and is, in fact, ln 2.674 (obtained by using the tables in reverse as previously explained). Thus

$$7.8914 = 0.9836 + 6.9078$$
$$= \ln 2.674 + \ln (10^3)$$
$$= \ln (2.674 \times 10^3)$$
$$= \ln 2674$$
$$\text{Antilog } 7.8914 = 2674$$

(c) This time examine the small table giving natural logarithms of 10^{-n}. These are required because the natural logarithm to be subtracted must contain a bar number such that the subtraction yields a natural logarithm in the range 0 to 2.3026.

The natural logarithm of 10^{-2} is $\bar{5}.3948$ and subtracting this from $\bar{6}.5132$ gives

$$(-6 + 0.5132) - (-5 + 0.3948)$$

which is

$$-1 + 0.1184 \qquad \text{that is, } \bar{1}.1184$$

which lies outside the range 0 to 2.3026.

The natural logarithm of 10^{-3} is $\bar{7}.0922$ and subtracting this from $\bar{6}.5132$ gives

$$(-6 + 0.5132) - (-7 + 0.0922)$$

which is

$$1 + 0.421 \qquad \text{that is, } 1.421$$

which lies within the range 0 to 2.3026 and is, in fact, by using the tables, the natural logarithm of 4.141.

Note that on this occasion the required mean difference of 3 is not available in the mean difference columns and we have a choice between 2 and 5 which give natural logarithms of 1.4209 and 1.4212 respectively. However, the 3 that we require to be added to 1.4207 to give 1.421 is closer to 2 than to 5 so the last digit of the number we are trying to determine is probably more accurately taken as 1 than 2.

Example 1.18

The voltage V_t across a certain component in an electrical circuit t seconds after the supply is disconnected is given by $V_t = 100e^{-0.1t}$. (a) Find the value of V_t after (i) 5.5 seconds   (ii) 45.5 seconds.   (b) After what time does $V_t = 50$ volts?

(a) (i) After 5.5 seconds $V_t = 100\ e^{-0.1 \times 5.5}$ that is

$$V_t = 100\ e^{-0.55}$$

Taking natural logarithms of both sides

$$\begin{aligned} \ln V_t &= \ln 100 + \ln(e^{-0.55}) \\ &= \ln 100 - 0.55 \ln e \\ &= 4.6052 - 0.55 \ (\text{since } \ln e = 1) \\ &= 4.0552 \end{aligned}$$

This lies outside the tables so $\ln V_t$ cannot be antilogged directly. Using the previous examples

$$\begin{aligned} \ln V_t &= 4.0552 \\ &= 2.3026 + 1.7526 \\ &= \ln 10 + \ln 5.769 \\ &= \ln(10 \times 5.769) \end{aligned}$$

therefore

$$\begin{aligned} V_t &= 10 \times 5.769 \\ &= 57.69 \text{ volts} \end{aligned}$$

The logarithm 2.3026 is obtained by taking the natural logarithm of a power of ten nearest to 4.0552. The logarithm 1.7526 is the logarithm which must be added to 2.3026 to give 4.0552, that is, it is obtained by subtracting 2.3026 from 4.0552.

(ii) After 45.5 seconds $V_t = 100\ e^{-0.1 \times 45.5}$

$$= 100\ e^{-4.55}$$

therefore

$$\ln V_t = \ln 100 - 4.55 \ln e$$
$$= 4.6052 - 4.55$$
$$= 0.0552$$
$$V_t = 1.057 \text{ volts}$$

Note that in antilogging 0.0552 the nearest natural logarithm is 0.0488 (being the natural logarithm of 1.05). The required mean difference is 64 and the figures available in the mean difference columns (columns 6 and 7 respectively) are 57 and 67. Since 67 is nearer to 64 than 57 the mean difference is taken as 7 rather than 6 and 1.05 becomes 1.057 to three places of decimals.

(b) When V_t = 50 volts

$$50 = 100 \, e^{-0.1t}$$

Taking natural logarithms

$$\ln 50 = \ln 100 - 0.1t \ln e$$

so that

$$0.1t = \ln 100 - \ln 50 \text{ (since } \ln e = 1)$$
$$= 4.6052 - (2.3026 + 1.6094)$$

where (2.3026 + 1.6094) is ln 50, that is, $\ln (10 \times 5)$ or $\ln 10 + \ln 5$. Thus

$$0.1t = 0.6932$$

and

$$t = 6.932 \text{ seconds}$$

(*Note* There is no need for antilogging in this case.)

Example 1.19

Find the value of e raised to the power e.

Let the answer be x so that $x = e^e$ and therefore

$$\ln x = e \ln e$$
$$= e \text{ (since } \ln e = 1)$$
$$= 2.7183$$

therefore

$$\ln x = 2.3026 + 0.4157$$
$$= \ln 10 + \ln 1.516 \qquad \text{(from tables)}$$
$$= \ln (10 \times 1.516)$$

therefore

$$x = 10 \times 1.516$$
$$= 15.16$$

and thus e^e $\qquad = 15.16$

Logarithms to any base in terms of common logarithms

Suppose the logarithm of any number N to any base a is x, that is

$$\log_a N = x$$

so that

$$a^x = N \quad \text{(by definition)}$$

and suppose the logarithm of a to the base 10 is y, that is

$$\log_{10} a = y$$

that is

$$10^y = a$$

then, since $a^x = N$ and $a = 10^y$

$$(10^y)^x = N \quad \text{(replacing } a \text{ by } 10^y \text{ in } a^x = N)$$

that is

$$10^{xy} = N$$

Taking common logarithms

$$xy \log_{10} 10 = \log_{10} N$$

and, since $x = \log_a N$ and $y = \log_{10} a$, then

$$\log_a N \times \log_{10} a = \log_{10} N$$

or

$$\log_a N = \log_{10} N \div \log_{10} a$$

In general terms, the logarithm of any number to any base is equal to the common logarithm of the number divided by the common logarithm of the base.

Example 1.20

Evaluate the following: (a) $\log_3 6$ (b) $\log_4 53$ (c) $\log_{2.3} 4.2$ (d) $\log_2 0.75$.

(a)
$$\log_3 6 = \frac{\log_{10} 6}{\log_{10} 3}$$

$$= \frac{0.7782}{0.4771}$$

$$= 1.6311$$

(b)
$$\log_4 53 = \frac{\log_{10} 53}{\log_{10} 4}$$

$$= \frac{1.7243}{0.6021}$$

$$= 2.8638$$

(c)
$$\log_{2.3} 4.2 = \frac{\log_{10} 4.2}{\log_{10} 2.3}$$

$$= \frac{0.6232}{0.3617}$$

$$= 1.7230$$

(d)
$$\log_2 0.75 = \frac{\log_{10} 0.75}{\log_{10} 2}$$

$$= \frac{\bar{1}.8751}{0.3010}$$

Converting to an all negative numerator

$$\log_2 0.75 = -\frac{0.1249}{0.3010}$$

$$= -0.4150$$

$$= -1 + 0.5850$$

$$= \bar{1}.5850 \text{ (using the familiar bar-number system)}$$

Example 1.21

Find without using tables of natural logarithms (a) ln 2.715   (b) ln 23.652.

(a)
$$\ln 2.715 = \log_e 2.715$$
$$= \frac{\log_{10} 2.715}{\log_{10} e}$$
$$= \frac{0.4338}{0.4343}$$
$$= 0.9988$$

(b)
$$\ln 23.652 = \frac{\log_{10} 23.652}{\log_{10} e}$$
$$= \frac{1.3738}{0.4343}$$
$$= 3.1633$$

Part (b) of this example offers an alternative method of finding natural logarithms of numbers outside the range 1 to 10. The other method using tables was considered earlier and was demonstrated in example 1.18.

FURTHER USE OF THE SLIDE RULE

The basic principles of the slide rule were presented in *First-year Technician Mathematics* and the methods used for multiplication and division were also considered. In this book we shall examine methods of obtaining powers of e, powers of any base, logarithms to base e and reciprocals from a slide rule. Since methods of obtaining trigonometric functions from a slide rule depend on the particular rule in use, these will not be discussed and the reader is advised to study the explanatory leaflet that is provided with the slide rule.

Powers of e

The top and bottom scales on most slide rules give powers of e. When the cursor is aligned with any number N_1 on the top scale and a number x on the D scale (the scale immediately below the slide) then $N_1 = e^{0.1x}$. Similarly, for an alignment of the cursor with any number N_2 on the bottom scale and a number x on the D scale then $N_2 = e^x$.

As an example consider the cursor at 3 on the D scale. The number on the top scale is 1.35 and the number on the bottom scale is 20. Thus

$$e^{0.3} = 1.35 \text{ and } e^3 = 20$$

Example 1.22

Evaluate (a) $e^{1.5}$ (b) $e^{0.5}$.

(a) Align cursor with 1.5 on D scale; read bottom scale to give 4.48, thus

$$e^{1.5} = 4.48$$

(b) Align cursor with 5 on D scale; read *top* scale to give 1.65, thus

$$e^{0.5} = 1.65$$

Example 1.23

If $e^x = 23.5$ find x.

Align cursor with 23.5 on *bottom* scale; read D scale to give 3.16, thus

$$e^{3.16} = 23.5 \text{ and } x = 3.16$$

Example 1.24

Evaluate $e^{10.6}$.

This may be obtained by writing $e^{10.6} = e^{5.3 \times 2}$, that is, $(e^{5.3})^2$. Align cursor with 5.3 on scale D; read bottom scale to give 200, thus

$$e^{5.3} = 200 \text{ and } (e^{5.3})^2 = 40\,000$$

In examples where the power of e is not a number having a square easily obtained the slide rule may be used for multiplication as shown below.

Example 1.25

Evaluate $e^{10.2}$.

Write $e^{10.2}$ as $(e^{5.1})^2$. Using the slide rule $e^{5.1} = 164$. Also using the slide rule as explained in *First-year Technician Mathematics* $164 \times 164 = 26900$ (an exact figure, unobtainable from a slide rule, is 26896). Thus

$$e^{10.2} = 26\,900$$

Natural logarithms

These may be obtained by using the above process in reverse, since if $e^x = N$ then $x = \log_e N$, that is, $\ln N$.

Example 1.26

Find the natural logarithms of (a) 7   (b) 50.

(a) Align cursor with 7 on bottom scale; read D scale to give 1.945, thus $e^{1.945} = 7$ and $\ln 7 = 1.945$.

(b) Align cursor with 50 on bottom scale; read D scale to give 3.91, thus $\ln 50 = 3.91$.

Note Although this process is easier than that using tables, the answer, however, is less accurate.

Powers of any number

The method is demonstrated below.

Example 1.27

Find $12.3^{1.2}$.

Align 1 on the C scale (the lowest of the scales on the slide) with 12.3 on the bottom (exponential) scale; move cursor to 1.2 on the C scale; read off bottom scale to give 20.4, thus $12.3^{1.2} = 20.4$.

Example 1.28

Evaluate $3.65^{2.4}$.

Align 1 on C scale with 3.65 on bottom scale; move cursor to 2.4 on C scale; read off bottom scale to give 22.4.

Reciprocals

The reciprocal of any number on the C scale is given on the centre scale on the slide (reading 1, 2, 3, etc. from *right* to *left*).

Example 1.29

Evaluate (a) 1/3.5   (b) 1/43.5   (c) 1/0.231.

(a) Align cursor with 35 on C scale; read centre scale to give 286 (take care here since the centre scale increases from right to left the figures of the answer are thus 286. To determine the position of the decimal point write 286 as 2.86 and consider what multiplier is required in the numerator of the problem to give this answer. The multiplier is 10 since 10/3.5 would give 2.86. The problem, however, is 1/3.5 that is, *one-tenth* of 10/3.5. The answer is thus one-tenth of 2.86 that is, 0.286.

(b) Align cursor with 435 on C scale; read off 23 on centre scale. Write 23 as 2.3. 100/43.5 would give 2.3, thus

$$1/43.5 \text{ would give } 2.3/100 \text{ that is, } 0.023.$$

(c) Align cursor with 231 on C scale; read off 433 centre scale. Write 433 as 4.33. 1/0.231 would give 4.33.

Exercise 1.3

(1) Find using tables the natural logarithms of (a) 2.341   (b) 23.41   (c) 234.1 (d) 2341.

(2) Find using common logarithm tables the natural logarithms of (a) 2.341 (b) 23.41   (c) 234.1   (d) 2341.

(3) The current i flowing in a circuit t seconds after connecting it to a supply is given by $i = 5(1 - e^{10t})$. Find (a) the current flowing after 0.16 seconds (b) the time after which the current is 4.7 A.

(4) The voltage V across a component t seconds after a d.c. supply is connected is given by $V = 50(1 - e^{t/100})$. Calculate the voltage (a) 10 seconds (b) 24.7 seconds after connection. After what time is the voltage equal to 25 V?

(5) If $4x = e^{5y}$ calculate y when $x = 27.18$.

(6) If $x^{0.4}/e^{-0.2y} = 5$ find (a) x when $y = 3.1$ (b) y when $x = 5.9$.

(7) Evaluate (a) $\log_3 24$ (b) $\log_2 17.61$ (c) $\log_5 198.3$.

(8) Evaluate (a) $\log_{2.3} 17.59$ (b) $\log_{0.12} 18.6$.

(9) Use a slide rule to evaluate (a) $e^{0.1x}$ when $x = 7.42$ (b) e^x when $x = 4.57$.

(10) Find the reciprocals of (a) 0.142 (b) 1.42 (c) 14.2 (d) 142 without using tables.

(11) Find the value of (a) $3.74^{2.4}$ (b) $6.78^{-1.8}$ without using tables.

(12) Use a slide rule to evaluate $\left(\dfrac{0.123 \times 11.4^{2.7}}{16.9^{3.1} \times 7.2}\right)^{-1}$

2 Algebra

Raising a number to a power or index (plural — indices) was discussed fully in the first volume of this series. The laws governing the manipulation of indices were also given in that volume and were considered again in chapter 1 of this volume, in so far as they applied to logarithms. In this section, further consideration will be given to manipulation of algebraic quantities raised to powers and some more difficult problems will be given. First, however, the basic definitions and rules are summarised below.

Summary of definitions and rules

(1) A^2, where A may represent any quantity, means $A \times A$; A^3 means $A \times A \times A$; and in general A^n means A multiplied by itself $(n-1)$ times, the base A appearing n times when A^n is written out in full.

(2) A^{-2} means the reciprocal of A^2, that is, $1/A^2$; A^{-3} means $1/A^3$; and in general A^{-n} means $1/A^n$.

(3) $A^2 \times A^3$ equals A^{2+3}, that is, A^5; $A^{-2} \times A^3$ equals A^{-2+3}, that is A^1 or A; and in general $A^n \times A^m = A^{m+n}$ where due consideration is given to the *signs* of m and n, that is, the addition of m and n is *algebraic*.

Note that the addition of powers applies only when numbers containing the same base (A in the examples given) are being multiplied. For example, $A^n \times B^m$ cannot be further simplified.

(4) $A^7 \div A^5 = A^{7-5}$ that is, A^2; $A^{-7} \div A^5 = A^{-7-5}$ that is, A^{-12} (which is $1/A^{12}$); $A^7 \div A^{-5} = A^{7-(-5)}$ that is, A^{7+5} or A^{12}; and in general $A^m \div A^n = A^{m-n}$ where due consideration is given to the signs of m and n. As before, the rule applies only when the numbers or quantities involved in the division are to the same base.

(5) $(A^4)^3 = A^{4 \times 3}$ that is, A^{12}; $(A^{-4})^3 = A^{-4 \times 3}$ that is, A^{-12}; $(A^{-4})^{-3} = A^{-4 \times (-3)} = A^{12}$; and in general $(A^m)^n = A^{mn}$ giving due consideration to the signs of m and n.

(6) $A^{1/2}$ means the number which when raised to the power 2 equals A ($A^{1/2}$ is the square root of A). $A^{1/3}$ means the number which when raised to the power 3 equals A ($A^{1/3}$ is the cube root of A). $A^{1/4}$ means the number which when raised to the power 4 equals A ($A^{1/4}$ is the fourth root of A). In general $A^{1/n}$ means the number which when raised to the power n equals A ($A^{1/n}$ is the nth root of A).

(7) Since $A^{2/3} = (A^2)^{1/3}$ by rule 5 then $A^{2/3}$ means the cube root of A^2 from rule 6. Similarly $A^{6/7}$ is the seventh root of A^6. Also $A^{-4/5}$ is the reciprocal of the fifth root of A^4, or alternatively the fifth root of A^{-4} and in general $A^{m/n}$ means the nth root of A^m.

Note that rules 2, 3 and 4 apply when the powers are integers (whole numbers) or fractions.

Example 2.1

If $a^x = b$ and $b^y = c$ find c in terms of a.

Since $c = b^y$ and $b = a^x$, then

$$c = (a^x)^y$$
$$= a^{xy} \quad \text{(rule 5)}$$

Example 2.2

If $a^{2/3} = b$ and $b^4 = 64$ find a^4.

Since $b^4 = 64$ then $b = 64^{1/4}$
$$= (4 \times 16)^{1/4}$$
$$= 4^{1/4} \times 16^{1/4}$$
$$= (4^{1/2})^{1/2} \times 2$$
$$= \sqrt{2} \times 2$$

And since $a^{2/3} = b$ then $a = b^{3/2}$ (obtained by raising both sides of the equation to the power 3 then taking the square root of both sides). So that

$$a = (2\sqrt{2})^{3/2}$$

and thus

$$a^4 = [(2\sqrt{2})^{3/2}]^4$$
$$= (2\sqrt{2})^6$$
$$= 64 \times 8$$
$$a^4 = 512$$

Example 2.3

If $(x^{3/2})^{-4} = (y^{-5})^{1/2}$ find $y^{1/3}$ in terms of x.

Simplifying gives

$$x^{-6} = y^{-5/2}$$

Squaring both sides and raising each side to the power -1 gives

$$x^{12} = y^5$$

so that $y = x^{12/5}$ (taking the fifth root of each side) and thus

$$y^{1/3} = (x^{12/5})^{1/3}$$
$$= x^{4/5} \quad \text{(rule 5)}$$

Example 2.4

If $a^{1/4}\, b^{1/3}\, c^{1/8} = 16$ and $bc^{3/8} = 8$ find a.

Now $b^{1/3}\, c^{1/8} = 16/a^{1/4}$ (dividing through by $a^{1/4}$) and since

$$bc^{3/8} = (b^{1/3}\, c^{1/8})^3$$

then

$$bc^{3/8} = \left(\frac{16}{a^{1/4}}\right)^3$$

so that

$$8 = \left(\frac{16}{a^{1/4}}\right)^3$$

since $bc^{3/8} = 8$ (given). Thus

$$8 = \frac{16^3}{a^{3/4}}$$

and

$$a^{3/4} = \frac{16^3}{8}$$

therefore

$$a = \left(\frac{16^3}{8}\right)^{4/3}$$

(taking the cube root of each side and raising to the power 4) thus

$$a = \frac{16^4}{8^{4/3}}$$

Now $16 = 2 \times 8$ so that $16^4 = 2^4 \times 8^4$, thus

$$a = \frac{2^4 \times 8^4}{8^{4/3}}$$

$$= 2^4 \times 8^{8/3} \qquad \text{(rule 4)}$$

also $8 = 2^3$ so

$$a = 2^4 \times (2^3)^{8/3}$$

$$= 2^4 \times 2^8$$

$$= 2^{12}$$

$$a = 4096$$

Note that simplification is greatly aided if it is possible to reduce numbers to powers of the same base.

Example 2.5

If $a^{3/2} = (x^2 + y^2)^{-2/3}$ obtain an equation relating x to a and y.

The equation must have x on the LHS. The method of solution must first therefore extract x from the RHS. Before this can be done the power shown and the bracket containing variables x and y must be removed. First raise each side of the equation to the power $-3/2$ (the reciprocal of the power on the RHS)

$$a^{3/2\,(-3/2)} = (x^2 + y^2)^{-2/3\,(-3/2)}$$

thus

$$a^{-9/4} = x^2 + y^2$$

Subtracting y^2 from both sides

$$a^{-9/4} - y^2 = x^2$$

and taking the square root of both sides

$$x = (a^{-9/4} - y^2)^{1/2}$$

which is the required equation. This example is actually a problem on transposition of formulae, the equation in this case containing powers. Other transpositions are considered after exercise 2.1.

Exercise 2.1

(1) If $y = x^{2/3}$ and $x^{1/3} = z^{1/5}$ obtain an equation relating y to z.

(2) If $x^{4/5} = y^2$ and $y^{1/3} = 4$ find the value of x.

(3) If $(a^{1/5})^{-2} = (b^{3/4})^4$ find b in terms of a.

(4) Simplify $x^{1/4} y^{2/5} z^{4/3} \div x^{-1/4} y^{1/5} z$.

(5) If $(a^3 b^2)^4 \div (a^2 b)^{-3} = 15$ and $a = 4$ find the value of b.

(6) Simplify $\frac{1}{3}$ of $\frac{2}{3} (x^4 y^3 z^2 \times x^2 y z^2)^{1/2}$.

(7) If $p^{1/3} q^{2/3} = 16$ when $p = 27$ find the value of q.

(8) Simplify $\left(\dfrac{m^{2/3} n^{4/5} p^2}{m^{1/3} n^2 p^3} \div m^{1/5} n^2 p^{1/4} \right)^{1/2}$.

(9) If $(a^3 b^2)^{1/4} = 3$ find the value of $a^{4/3} b^{1/3}$.

(10) Simplify $(r^2 s^{1/2} t^{-1/4} \div r^{1/3} s\, t^{1/2}) \times (r s\, t)^{-1/2}$.

TRANSPOSITION OF FORMULAE

Transposition of formulae involves rearranging equations so that a particular variable is expressed in terms of other variables. Transposition was discussed in some detail in volume 1 of this series; in the present volume no further rules are given but more difficult problems and examples together with the basic rules are given below.

In all transpositions it must be remembered that an equation is only being rearranged; an equation is a statement of equality between two expressions or numbers and as such must not be changed in any way that renders the equation untrue. In practical terms this means that any mathematical process applied to one side of an equation must be applied in exactly the same way to the other side of the equation. Thus if the LHS is multiplied or divided by a number or term, then the RHS must be multiplied or divided respectively by the same number or term. Similarly if the LHS is raised to any power, the RHS must be raised to the same power. Also if any number or term is added to or subtracted from the LHS the same number or term must be added to or subtracted from the RHS. This rule is extremely important; neglect of it must lead to error.

In obtaining any variable in terms of other variables the following basic sequence should be employed. First extract the term or terms containing the variable to be expressed in terms of others (the subject). If the subject lies within a bracket and the bracketed terms are raised to a power, this bracket must first be removed by raising both sides of the equation to the reciprocal of the power concerned; for example, if the bracketed terms are raised to the power $\frac{1}{2}$ then both sides should be squared (raised to power 2). Once extracted the terms containing the subject should be factorised*, if necessary, to extract the subject. Terms not containing the subject should then be removed from the side of the equation containing the subject, with the ultimate aim of isolating the subject on one side of the equation. The following examples illustrate the general approach to transposition problems.

Example 2.6

If $s = ut + \frac{1}{2} ft^2$ express f in terms of s, u and t.

To extract the term containing f subtract ut from both sides to give

$$\tfrac{1}{2} ft^2 = s - ut$$

Dividing through by t^2 and multiplying through by 2 gives

$$f = 2\left(\frac{s - ut}{t^2}\right)$$

which is the required relationship.

Example 2.7

The current I in a series circuit connected across an alternating voltage supply of V volts is given by

$$I = \frac{V}{\sqrt{\{R^2 + [\omega L + 1/(\omega C)]^2\}}}$$

---

*Factorisation of an algebraic expression is the determination of further expressions or terms which, when multiplied together yield the original expression. It was dealt with in volume 1 and is considered again later in this chapter.

where R, L and C represent the circuit resistance in ohms, inductance in henrys and capacitance in farads respectively and ω is the angular velocity. Obtain an equation for R in terms of the remaining variables.

First extract the terms containing or associated with the subject variable R. (R is in fact in the denominator of the RHS but cannot be immediately extracted since it is associated with other terms, the expression as a whole being raised to the power $\frac{1}{2}$.) Multiply both sides by this expression $\sqrt{\{R^2 + [\omega L + 1/(\omega C)]^2\}}$ to give

$$I\sqrt{\{R^2 + [\omega L + 1/(\omega C)]^2\}} = V$$

Now divide through by I

$$\sqrt{\{R^2 + [\omega L + 1/(\omega C)]^2\}} = \frac{V}{I}$$

Square both sides

$$R^2 + [\omega L + 1/(\omega C)]^2 = \left(\frac{V}{I}\right)^2$$

Subtract $[\omega L + 1/(\omega C)]^2$ from both sides

$$R^2 = (V/I)^2 - [\omega L + 1/(\omega C)]^2$$

Take the square root of both sides

$$R = \sqrt{\{(V/I)^2 - [\omega L + 1/(\omega C)]^2\}}$$

which is the required expression.

Example 2.8

If $Z = \sqrt{[R^2 + (X_1 - X_2)^2]}$ rearrange the formula to obtain an expression for X_1 in terms of Z, R and X_2.

The required subject is associated with other terms in a bracket raised to the power $\frac{1}{2}$. First, square both sides to give

$$Z^2 = R^2 + (X_1 - X_2)^2$$

Now remove R^2 from both sides to give

$$Z^2 - R^2 = (X_1 - X_2)^2$$

The subject is still within a bracket raised to the power 2; take the square root of both sides (that is, raise both sides to the power $\frac{1}{2}$, being the reciprocal of 2) to give

$$\sqrt{(Z^2 - R^2)} = X_1 - X_2$$

Now add X_2 to both sides

$$\sqrt{(Z^2 - R^2)} + X_2 = X_1$$

which is the required expression.

Example 2.9

The voltage gain A_v of a triode amplifier is given by

$$A_v = \frac{\mu R_L}{r_a + R_L}$$

where μ is the valve amplification factor

$\quad R_L$ is the load resistance

$\quad r_a$ is the valve anode resistance

Obtain an expression for R_L in terms of A_v, μ and r_a.

The subject R_L is contained both in the numerator and denominator of the RHS. The transposition is made much easier if R_L is contained only in the numerator. Multiply both sides by $(r_a + R_L)$ to give

$$A_v r_a + A_v R_L = \mu R_L$$

Now collect all terms containing R_L by subtracting μR_L from both sides to give

$$A_v r_a + A_v R_L - \mu R_L = 0$$

Put the term not containing R_L on the RHS by subtracting $A_v r_a$ from both sides

$$A_v R_L - \mu R_L = -A_v r_a$$

Factorise the LHS

$$R_L (A_v - \mu) = -A_v r_a$$

Now divide both sides by $(A_v - \mu)$ to give

$$R_L = \frac{-A_v r_a}{A_v - \mu}$$

which is the required expression.

As shown in the foregoing examples it is essential not to do anything to the equation which renders it untrue and to bear in mind that the subject or terms containing the subject cannot be extracted from a bracketed expression until the bracket is removed. This is usually effected by raising to a power or multiplying through. The remaining example shows the use of logarithms in extracting the subject when the subject itself is a power.

Example 2.10

If $I = ad^n$ obtain n in terms of I, a and d.

Taking the nth root of both sides, which may appear a reasonable thing to do, will not in fact produce an equation with n alone on one side. The equation produced would be $I^{1/n} = a^{1/n} d$. The method always used where the required subject is a power is to take logarithms since $\log x^2 = 2 \log x$, and $\log x^3 = 3 \log x$, then $\log x^n = n \log x$ which brings the original power down into the equation. For the equation given, $I = ad^n$, take logs of both sides

$$\log I = \log a + \log d^n$$

(Note that the multiplication sign implied between a and d^n changes to addition on taking logs.) Thus

$$\log I = \log a + n \times \log d$$

Removing $\log a$ from both sides

$$\log I - \log a = n \log d$$

or, since $\log I - \log a = \log (I/a)$

then

$$\log (I/a) = n \log d$$

Dividing through by $\log d$ gives

$$n = \frac{\log (I/a)}{\log d}$$

which is the required expression.

Exercise 2.2

(1) If $f_r = 1/2\pi \sqrt{(LC)}$ determine an equation for C in terms of f_r, π and L.

(2) The input resistance R_{IN} of a transistor amplifier is given by

$$R_{IN} = h_i - \frac{h_r h_f}{G_L + h_o}$$

where h_i, h_r, h_f, h_o are the transistor hybrid parameters and G_L is the amplifier load conductance. Obtain an equation for G_L in terms of the remaining variables.

(3) The output impedance Z_{od} of an amplifier using a field-effect transistor is given by

$$Z_{od} = \frac{R'_L}{1 + g_{fs}R'_L}$$

where g_{fs} is a transistor parameter and R'_L is the effective load resistance. Obtain an expression for R'_L in terms of Z_{od} and g_{fs}.

(4) The h parameters of a transistor are related by the equation

$$h_{oc} = \frac{h_{ob}}{1 + h_{fb}}$$

Express h_{fb} in terms of h_{oc} and h_{ob}.

(5) The following equation gives the torque T of a d.c. machine in terms of the number of poles p, the number of conductors Z, the magnetic flux Φ, a machine constant a and the armature current I_a

$$T = \frac{p Z \Phi}{\pi a} I_a$$

Make Φ the subject of the equation.

(6) The voltage across the resistance V_e in a circuit consisting of a resistor and inductor is given by

$$V_e = E(1 - e^{-\tau t})$$

where E is the d.c. supply voltage

   t is time in seconds measured from switching on

   τ is the time constant

   e is a constant

By using logarithms, (a) to the base 10, (b) to the base e, express τ in terms of E, t and e.

(7) If $X_L^2 = (L/C) - R_L^2$ express C in terms of X_L, L and R_L.

(8) If

$$f_r = \frac{1}{2\pi} \left(\frac{1}{LC} - \frac{R_L^2}{L^2} \right)^{1/2}$$

find an equation relating R_L to the remaining variables.

(9) The no-load power factor of a transformer, symbol $\cos \theta_0$, is given by

$$\cos \theta_0 = \frac{I_c}{(I_c^2 + I_m^2)^{1/2}}$$

where I_c and I_m are components of the no-load current. Find I_m in terms of $\cos \theta_0$ and I_c.

(10) An equation connected with the reliability of a system using two electrical components is

$$m_p = \frac{1}{\lambda_1} + \frac{1}{\lambda_2} - \frac{1}{\lambda_1 + \lambda_2}$$

Find λ_2 in terms of λ_1 and m_p. (*Hint* Obtain a quadratic in λ_2 and solve for λ_2.)

SIMULTANEOUS LINEAR EQUATIONS

A linear equation with one variable is one in which the variable appears raised to a power no higher than 1. Linear equations and their solution were considered in volume 1 and, as shown there, the methods employed are not unlike those used in transposition of formulae, that is, the ultimate aim is to isolate the single variable on one side of the equation. Since it is the only variable in the equation, the remaining parts of the equation being numbers, the numerical solution follows.

Simultaneous linear equations are those equations containing one or more variables, (none of which are raised to a power higher than unity) which apply at the same time to a particular circuit or system. These too were considered in volume 1 and, as stated, two equations are required to find the numerical value of each of two variables which satisfy both equations; three equations are required when three variables are involved; and in general, n equations are required when n variables are involved. It should be noted that solving simultaneous linear equations determines the appropriate single numerical value that each variable must have for all the equations to be true at any one time.

One of the most common situations in which simultaneous linear equations occur is in the solution of circuit problems involving determination of circuit currents.

Example 2.11
Currents I_1 and I_2 in a certain network are related by the equations

$$3I_1 + 5I_2 = 10 \tag{1}$$

$$4I_1 - 2I_2 = 8 \tag{2}$$

Solve these simultaneous equations to obtain the values of I_1 and I_2.

The method to be employed is elimination (see *First-year Technician Mathematics*, chapter 4) in which one of the unknowns, I_1 or I_2 in this example, is eliminated to leave a single linear equation involving the other unknown. To eliminate I_2 multiply equation 1 by 2, equation 2 by 5 and add the new equations. Thus (equation 1) $\times$ 2 gives

$$6I_1 + 10I_2 = 20$$

(equation 2) $\times$ 5 gives

$$20I_1 - 10I_2 = 40$$

and addition gives

$$26I_1 = 60$$

thus

$$I_1 = \frac{60}{26}$$

$$= \frac{30}{13} \text{ A}$$

Substituting this value in equation 1

$$3 \times \frac{30}{13} + 5I_2 = 10$$

so that

$$5I_2 = 10 - \frac{3 \times 30}{13}$$

$$= \frac{40}{13}$$

and

$$I_2 = \frac{40}{5 \times 13}$$

$$= \frac{8}{13} \text{ A}$$

When I_1 was determined, equation 1 was used to find I_2, so I_2 should now be checked using equation 2.

LHS of equation 2

$$\frac{4 \times 30}{13} - \frac{2 \times 8}{13} = \frac{104}{13}$$

$$= 8$$

$$= \text{RHS of equation 2}$$

The values $I_1 = \frac{30}{13}$ A, $I_2 = \frac{8}{13}$ A satisfy equations 1 and 2 simultaneously.

Notice that in the elimination method the coefficients of the unknown to be eliminated are made the same in both equations by multiplying each by an appropriate factor. If the coefficient in one equation cannot be made equal to the coefficient in the other equation by use of a simple integer as a multiplier, then a useful method to remember is to multiply each equation throughout by the coefficient in the other equation of the unknown to be eliminated. This was used in the preceding example.

Example 2.12

Determine the values of V_1 and V_2 given that

$$4V_1 + 3V_2 = 100 \tag{1}$$

$$2V_1 + 6V_2 = 140 \tag{2}$$

In this example substitution will be used, where one unknown is expressed in terms of the other, using one of the equations, and the expression substituted in the other equation. Thus using equation 1 to find V_1 in terms of V_2, since

$$4V_1 + 3V_2 = 100$$

then
$$4V_1 = 100 - 3V_2 \quad \text{(subtracting } 3V_2 \text{ from both sides)}$$

and

$$V_1 = 25 - \tfrac{3}{4} V_2 \quad \text{(dividing through by 4)}$$

Substitute for V_1 in equation 2 to give a single linear equation in V_2
$$2(25 - \tfrac{3}{4} V_2) + 6V_2 = 140$$

Thus

$$50 - \tfrac{3}{2} V_2 + 6V_2 = 140 \quad \text{(multiplying out)}$$

and

$$\tfrac{9}{2} V_2 = 90 \quad \text{(collecting } V_2 \text{ terms and numerical terms)}$$

so that

$$V_2 = 20$$

thus from equation 1, since

$$V_1 = 25 - \tfrac{3}{4} V_2$$

then

$$V_1 = 25 - 15$$
$$= 10$$

The answer is thus $V_1 = 10$, $V_2 = 20$.

Substitution or elimination may be used as required. A different method was chosen for each of the preceding examples purely for demonstration purposes. Note also that simultaneous linear equations in two unknowns may also be solved graphically by plotting graphs of each equation and noting the point of intersection (since at this point the values of the variables satisfy both equations simultaneously).

Example 2.13

Apply Kirchhoff's laws to the circuit shown in figure 2.1 and determine the values of the currents shown by solving the resulting simultaneous equations.

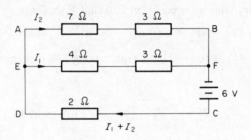

Figure 2.1

Kirchhoff's laws state first, that the algebraic sum of currents at any junction is zero, and second, that the algebraic sum of voltages around any closed mesh is zero. Alternatively these may be expressed as first, that the sum of the currents entering a junction equals the sum of the currents leaving the junction, and second, that in any closed mesh the sum of the electromotive forces (e.m.f.s) equals the sum of the potential differences (p.d.s). Both laws and their application are considered in detail in appropriate books on basic electrical theory and they are introduced here merely because their application inevitably produces simultaneous linear equations.

Kirchhoff's current law has already been applied to the circuit of figure 2.1 in that the current $I_1 + I_2$ entering the resistive branches, from the positive pole of the battery, is equal to the sum of the currents I_1 and I_2 leaving the resistive branches for the negative pole of the battery.

Applying Kirchhoff's voltage law to mesh CDEF

$$\text{sum of e.m.f.s} = 6$$

$$\text{sum of p.d.s} = 2(I_1 + I_2) + 4I_1 + 3I_1$$
$$= 9I_1 + 2I_2$$

Thus

$$6 = 9I_1 + 2I_2 \tag{1}$$

Applying Kirchhoff's voltage law to mesh CDAB

$$\text{sum of e.m.f.s} = 6$$

$$\text{sum of p.d.s} = 2(I_1 + I_2) + 7I_2 + 3I_2$$
$$= 2I_1 + 12I_2$$

Thus
$$6 = 2I_1 + 12I_2 \tag{2}$$

The problem is now purely mathematical and requires the solution of the simultaneous equations 1 and 2.

Multiply equation 1 throughout by 6 to give

$$36 = 54I_1 + 12I_2$$

Subtract equation 2 from this to give

$$36 - 6 = 54I_1 - 2I_1 + 12I_2 - 12I_2$$

that is

$$30 = 52I_1$$

and thus

$$I_1 = \frac{15}{26} \text{ A}$$

Substituting in equation 1 to find I_2

$$6 = \frac{9 \times 15}{26} + 2I_2$$

thus

$$2I_2 = 6 - \frac{9 \times 15}{26}$$
$$= \frac{21}{26}$$

and

$$I_2 = \frac{21}{52} \text{ A}$$

Expressing the solution decimally $I_1 = 0.577$ A and $I_2 = 0.403$ A.
   Check in equation 2 (equation 1 has been used to find I_2)

$$\text{LHS} = 6$$

$$\text{RHS} = 2 \times 0.577 + 12 \times 0.403$$

$$= 1.15 + 4.85 \quad \text{(to 2 places of decimals)}$$

$$= 6$$

The solution is correct. (Note that the use of a slide rule may give 5.99; only four or five-figure logarithms or an electronic or mechanical calculator would give exactly 6.)

Example 2.14

A 12 V d.c. supply of negligible internal resistance is used to charge two 6 V batteries via a resistor of value 0.5 Ω. If the internal resistances of the batteries are 0.2 Ω and 0.18 Ω calculate the current drawn by each battery from the supply.

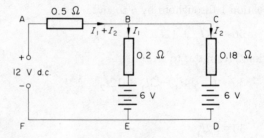

Figure 2.2

The circuit is shown in figure 2.2. Applying Kirchhoff's law to mesh ABEF

$$\text{sum of e.m.f.s} = 12 - 6 \text{ V}$$
$$= 6 \text{ V}$$

$$\text{sum of p.d.s} = 0.5\,(I_1 + I_2) + 0.2I_1$$
$$= 0.7I_1 + 0.5I_2$$

Thus

$$6 = 0.7I_1 + 0.5I_2 \tag{1}$$

Applying Kirchhoff's law to mesh ACDF

$$\text{sum of e.m.f.s} = 12 - 6 \text{ V}$$
$$= 6 \text{ V}$$

$$\text{sum of p.d.s} = 0.5\,(I_1 + I_2) + 0.18I_2$$
$$= 0.5I_1 + 0.68I_2$$

Thus $\qquad\qquad 6 \qquad = 0.5I_1 + 0.68I_2 \qquad\qquad$ (2)

Multiply equation 1 by 0.5

$$3 = 0.35\,I_1 + 0.25I_2 \qquad\qquad (3)$$

Multiply equation 2 by 0.7

$$4.2 = 0.35I_1 + 0.476I_2 \qquad\qquad (4)$$

Subtract equation 3 from equation 4

$$1.2 = 0.226I_2$$

thus

$$I_2 = \frac{1.2}{0.226}$$

$$= 5.31 \text{ A}$$

Substitute this value in equation 3

$$3 = 0.35I_1 + 0.25 \times 5.31$$

$$= 0.35I_1 + 1.327$$

thus

$$0.35I_1 = 3 - 1.327$$

$$= 1.673$$

$$I_1 = \frac{1.673}{0.35}$$

$$= 4.78 \text{ A}$$

It is left to the reader to check the solution using equation 4.

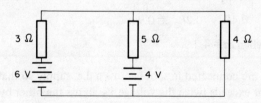

Figure 2.3

Exercise 2.3

(1) Find the current supplied by each battery in the circuit shown in figure 2.3 by using Kirchhoff's laws.

(2) A battery of internal resistance 0.5 Ω and open circuit e.m.f. 5 V is connected in parallel with a battery of internal resistance 0.4 Ω and open circuit e.m.f. 6 V. The combination is connected across a 10 Ω load resistor. Find the current in the load.

(3) Solve the following simultaneous equations for I_1 and I_2

   (a) $3I_1 + 4I_2 = 8$ (b) $0.3I_1 = 8 - 0.7I_2$
   $\quad\;\, 5I_1 - 3I_2 = 7$ $\quad\;\, 0.6I_1 = 6 + 0.5I_2$

(4) If three times the sum of two numbers is greater than twice their difference by 6 and one number is five times greater than the other, calculate the numbers.

(5) A train travels from town X to town Y and back again at an average speed of V km/h. The average speed of the journey from X to Y is $(V - 10)$ km/h and of the journey from Y to X is $(V + 12)$ km/h. Calculate the value of V if the times taken for the two journeys differ by half an hour.

(6) If $X = kM + c$ find k and c given that when $M = 7$, $X = 8$ and when $M = 5$, $X = 10$.

(7) Solve for a and b

$$\frac{1}{a} + \frac{2}{b} = \frac{3}{4}$$

$$\frac{4}{a} - \frac{7}{b} = \frac{1}{2}$$

(8) Two power levels P_1 and P_2 are related by the equations

$$0.18P_1 + 0.63P_2 = m$$

$$0.4P_1 - 0.2P_2 = 0.9m$$

Find P_1 and P_2 when $m = 4.2$.

(9) Two resistors are connected in series across a d.c. supply so that the voltage V_1 across one resistor exceeds twice the voltage V_2 across the other by 3 V and the sum of twice the voltage V_1 and four times the voltage V_2 is 15 V. If the current in the circuit is 2 A find the power absorbed by each resistor.

(10) The linear part of a diode anode characteristic is described by the equation

$$I_a = kV_a + c$$

where I_a is the anode current (mA), V_a is the anode voltage (V) and k, c are constants. For a certain diode when the anode current is 10 mA the anode voltage is 6.3 V and when the anode current is 20 mA the anode voltage is 14 V. Find the values of k and c for this diode.

QUADRATIC EQUATIONS

So far we have considered only algebraic expressions in which the variable is raised to a power no higher than unity, that is, linear expressions of the form $ax + b$ in which x is the variable and a and b are constants. When such an expression is given a value we obtain a linear equation and there is one value of the variable which satisfies the equation.

An expression of the form $ax^2 + bx + c$ in which x is the variable and a, b and c are constants is called a *quadratic* expression and when such an expression is given a numerical value the resulting *quadratic equation* is satisfied by either of *two* values of the variable.

Example 2.15

Two resistors, one of value 15 Ω, are connected in series across a supply voltage such that the p.d. across the unknown resistor is 10 V and the total power in the circuit is 5 W. Determine the equation relating the current i to the other given values.

Power in a resistance = voltage across resistance × current through it

or

= (current)^2 × resistance

Thus the total power in this circuit, 5 W, is equal to the power in the unknown resistor, that is $10i$ W (p.d. is 10 V), added to the power in the 15 Ω resistor which is $15i^2$ W. Thus

$$5 = 15i^2 + 10i$$

and the required equation is

$$15i^2 + 10i - 5 = 0$$

This equation is a quadratic equation and there are two possible values of current that would enable the given conditions to be established.

Example 2.16

Two resistors of value R_1 Ω and R_2 Ω are connected in series to give a total resistance of 10 Ω. When connected in parallel their total resistance is 2.4 Ω. Obtain an equation relating either R_1 or R_2 to the other given values.

Since

$$R_1 + R_2 = 10 \quad \text{(series)}$$

then

$$R_2 = 10 - R_1$$

Now

$$\frac{R_1 R_2}{R_1 + R_2} = 2.4 \quad \text{(parallel)}$$

so that

$$R_1 R_2 = 2.4\,(R_1 + R_2) \quad \text{[Multiplying through by } (R_1 + R_2)]$$

and, replacing R_2 by $10 - R_1$ and $(R_1 + R_2)$ by 10

$$R_1\,(10 - R_1) = 2.4 \times 10$$

that is

$$10R_1 - R_1^2 = 24$$

or, rearranging

$$R_1^2 - 10R_1 + 24 = 0$$

which is the required equation. (If an equation for R_2 is required instead replace R_1 by $(10 - R_2)$ instead of R_2 by $(10 - R_1)$ in the example given above).

Solution of quadratic equations

Solving quadratic equations means finding the values of the variable that satisfy the equation, that is, which make the equation true. These values are called the *roots* of the equation. There are three methods that may be used to solve quadratic equations (1) factorisation   (2) completing the square   (3) use of the formula. These three methods will now be considered in turn.

Factorisation

If two expressions each containing a variable raised to a power no greater than unity are multiplied together the resulting expression will contain the variable raised to a power no higher than two and this will be a quadratic expression. For example, consider $(x + 2)(x - 3)$. In expanding the brackets, that is, in carrying out the multiplication, each term in the first bracket, (x and 2), must be multiplied by each term in the second bracket, (x and -3). Thus

$$(x + 2)(x - 3) = (x + 2)x + (x + 2)(-3)$$
$$= x^2 + 2x - 3x - 6$$
$$= x^2 - x - 6$$

which is a quadratic expression. The bracketed expressions $(x + 2)$ and $(x - 3)$ are *factors* of the quadratic expression $x^2 - x - 6$.

Example 2.17
Solve the quadratic equation

$$x^2 - x - 6 = 0$$

From above

$$x^2 - x - 6 = (x + 2)(x - 3)$$

and so

$$(x + 2)(x - 3) = 0$$

If either bracketed expression is zero the LHS as a whole will be zero. Thus if

$$x + 2 = 0$$

giving

$$x = -2$$

or if

$$x - 3 = 0$$

giving

$$x = 3$$

then the LHS is zero. The values $x = -2$ and $x = 3$ are those that satisfy the equation, that is, they are the roots of $x^2 - x - 6 = 0$.

Any quadratic equation may be factorised into two expressions that give the roots of the equation using the method shown above. In general terms, provided the RHS of the equation is zero and the coefficient of x^2 is 1, as in example 2.17, then if x is the variable and R_1, R_2 are the roots, the factors are $(x - R_1)$ and $(x - R_2)$. This is readily seen by equating the product of these two factors to zero.

$$(x - R_1)(x - R_2) = 0$$

Then for the left-hand bracketed expression to be zero $x = R_1$ and for the right-hand bracketed expression to be zero $x = R_2$. Thus R_1, R_2 are the roots of the equation.

By examination of the way in which the factors are multiplied together to give the orginial quadratic we can obtain some guide as to how the values of R_1 and R_2 may be obtained by inspection. Consider the equation

$(x - R_1)(x - R_2) = 0$ where R_1, R_2 are the roots of the equation. Multiplying out we obtain

$$(x - R_1)x + (x - R_1)(-R_2) = 0$$

so that

$$x^2 - R_1x - R_2x + R_1R_2 = 0$$

that is

$$x^2 - (R_1 + R_2)x + R_1R_2 = 0$$

In looking for the values of R_1 and R_2 when factorising any quadratic we see therefore that the algebraic sum of these values must equal the coefficient of x in the equation to be factorised, and that the product of these same values must equal the constant in the original equation. For clarification of this general statement look again at example 2.17 in which $x^2 - x - 6$ is factorised into $(x - 3)(x + 2)$. The numbers in the factors, that is, -3 and 2, are such that their (algebraic) sum is $-3 + 2$, that is, -1, which is the coefficient of x in the quadratic expression $x^2 - x - 6$; and their product is $(-3) \times (+2)$, that is, -6, which is the constant in the expression $x^2 - x - 6$.

Example 2.18
Find the roots of the equation $x^2 + 6x + 5 = 0$.

To find the factors we require two numbers which when multiplied together give 5 (the constant) and when added give 6 (the coefficient of x). Two such numbers are 5 and 1 to give

$$(x + 5)(x + 1) = 0$$

and the roots are -5 (to make $x + 5$ zero) and -1 (to make $x + 1$ zero). Note that the roots have the opposite sign to the numbers in the factors.

Example 2.19
Solve $4x^2 + 5x - 6 = 0$.

To ease the problem of factorisation reduce the coefficient of x to unity by dividing throughout by 4; that is

$$x^2 + \frac{5}{4}x - \frac{6}{4} = 0$$

The numbers in the factors when multiplied must give $-6/4$ and when added must give $5/4$. Two such numbers are $8/4$ and $-3/4$ which when multiplied give $(8/4) \times (-3/4)$ that is, $-24/16$ or $-6/4$ and when added give $(8/4) - (3/4)$ that is, $5/4$ as required. Thus

$$x^2 + \frac{5}{4}x - \frac{6}{4} = 0$$

may be written

$$(x + \tfrac{8}{4})(x - \tfrac{3}{4}) = 0$$

or

$$(x + 2)(x - \tfrac{3}{4}) = 0$$

giving the roots as -2 and $\frac{3}{4}$. Note that $(x + 2)(x - \frac{3}{4})$ can also be written as $(x + 2)(4x - 3)$, obtained by multiplying the second bracketed expression by 4, and $(x + 2)(4x - 3) = 0$ expands to give $4x^2 + 5x - 6 = 0$ which is the original equation. The roots of course are the same, to make $(x + 2)$ zero, $x = -2$, to make $4x - 3$ zero, $4x = 3$ or $x = \frac{3}{4}$ as before.

Example 2.20

Find the possible values of two resistors which in series have a total resistance of $18 \, \Omega$ and in parallel have a total resistance of $4 \, \Omega$.

Let the resistance of one resistor be $R \, \Omega$ so that the resistance of the other is $(18 - R) \, \Omega$. The resistance of the parallel combination is

$$\frac{R(18 - R)}{18} \, \Omega$$

which equals 4 (given). Thus

$$\frac{R(18 - R)}{18} = 4$$

and

$$18R - R^2 = 72 \qquad \text{(expanding the LHS and multiplying through by 18)}$$

Rearranging

$$R^2 - 18R + 72 = 0$$

Two numbers that multiply to give 72 and add to give -18 are -12 and -6. The equation factorises to give

$$(R - 12)(R - 6) = 0$$

and possible values of R are 12 Ω and 6 Ω. The possible values of the other resistor are 6 Ω and 12 Ω respectively.

Example 2.21

Find the values of x which satisfy the equation

$$3x^2 + 12x - 4 = x^2 + 10$$

Gathering like terms

$$3x^2 - x^2 + 12x - 4 - 10 = 0$$

that is

$$2x^2 + 12x - 14 = 0$$

Dividing through by 2

$$x^2 + 6x - 7 = 0$$

Factorising

$$(x - 1)(x + 7) = 0$$

The values of x are 1 and -7.

   In the examples using factorisation to solve quadratic equations it was not unduly difficult to obtain the factors and thus the roots of the equation. In many quadratic equations the factors are not obvious and other methods have to be employed. The main method is by use of a formula, to be considered later, which is derived from the method now to be considered, that of completing the square.

Completing the square

Some quadratic expressions may be factorised to give two factors that are exactly the same; in other words the expression is a perfect square. Some examples are $x^2 + 2x + 1$ which equals $(x + 1)^2$, $x^2 + 4x + 4$ which equals $(x + 2)^2$, $x^2 - 6x + 9$ which equals $(x - 3)^2$, and so on.

   If the perfect square forms part of a quadratic equation in which the other side is zero the implication is that the roots are equal; that is, in $x^2 + 2x + 1 = 0$, which may be written $(x + 1)(x + 1) = 0$, the roots are -1 and -1. The significance of such a solution is considered in chapter 3.

Where one side of the quadratic equation is not a perfect square, terms may be added to both sides to make it so and the roots may then be found as shown in the following example.

Example 2.22

Solve the equation $x^2 + 4x + 2 = 0$ by completing the square.

As indicated above a perfect square is $x^2 + 4x + 4$ which is the square of $x + 2$. Add 2 to *both* sides of the equation to give

$$x^2 + 4x + 4 = 2$$

and thus

$$(x + 2)^2 = 2$$

Taking the square root of each side

$$x + 2 = \pm \sqrt{2}$$

(Note that the square root of 2 may be $+ \sqrt{2}$ or $- \sqrt{2}$) and so

$$x = -2 \pm \sqrt{2}$$
$$= -2 + \sqrt{2} \text{ or } -2 - \sqrt{2}$$
$$= -0.586 \text{ or } -3.414$$

The roots of the equation are thus $x = -0.586$ and -3.414 and, clearly, factors such as $(x + 0.586)$ and $(x + 3.414)$ which are the factors of $(x^2 + 4x + 2)$ are certainly not easily obtained by inspection as were the factors in preceding examples.

Example 2.23

Solve the equation $x^2 + 8x + 10 = 0$ by completing the square.

The LHS will be made a perfect square by adding 6 to it to give $x^2 + 8x + 16$, that is, $(x + 4)^2$. Add 6 to *both* sides

$$x^2 + 8x + 16 = 6$$
$$(x + 4)^2 = 6$$
$$x + 4 = \pm \sqrt{6}$$
$$x = -4 + \sqrt{6} \text{ and } -4 - \sqrt{6}$$
$$= -1.55 \text{ and } -6.45$$

which are the required roots.

The question that naturally arises is how does one know what to add to each side of the equation to obtain a perfect square on one side? Examination of what constitutes a perfect square will lead to a method of determining this.

The square of any term such as $(x + a)$ is obtained by squaring the first term to give x^2, multiplying both terms, a and x, together and doubling to give $2ax$, and squaring the last term to give a^2, that is

$$(x + a)^2 = x^2 + 2ax + a^2$$

It will be seen that the coefficient of x in the expanded expression is 2 times the second term in the expression being squared. Alternatively, if $x^2 + 2ax$ are the first two terms of the expanded expression then a will be the second term in the expression being squared; that is, the second term in the expression being squared is one half of the coefficient of x in the expanded expression. Thus, in example 2.23 above, since $x^2 + 8x$ are the first terms in the expanded expression then $[x + (8/2)]$ that is, $(x + 4)$ is the expression to be squared. The term to be added to each side is then the difference between the expanded expression which we have to solve, $x^2 + 8x + 10$ in this case, and $(x + 4)^2$, that is $x^2 + 8x + 16$, the expression required to find the solution; a difference of 6 in this example.

Study the following examples carefully.

Example 2.24
Solve $x^2 + 10x + 15 = 0$ by completing the square.

Half the coefficient of x is 5, thus $(x + 5)^2$ is required to solve the equation. Now $(x + 5)^2 = x^2 + 10x + 25$ and we have $x^2 + 10x + 15$ so the difference is $25 - 15 = 10$. Add 10 to both sides of the equation

$$x^2 + 10x + 25 = 10$$

thus

$$(x + 5)^2 = 10$$

and

$$x + 5 = \pm\sqrt{10}$$

so that

$$x = -5 \pm \sqrt{10}$$

which are the roots of the equation as required.

Example 2.25
Solve $4x^2 + 16x + 5 = 0$.

First, divide through by 4 to give a coefficient of x equal to unity. This is necessary in order to apply the method of halving the coefficient of x. Dividing by 4 gives

$$x^2 + 4x + \frac{5}{4} = 0$$

Half the coefficient of x is 4/2, that is 2, so the required square is $(x + 2)$. Now

$$(x + 2)^2 = x^2 + 4x + 4$$

and we have

$$x^2 + 4x + \frac{5}{4}$$

thus the difference is

$$4 - \frac{5}{4} = \frac{11}{4}$$

Add 11/4 to both sides

$$x^2 + 4x + \frac{5}{4} + \frac{11}{4} = \frac{11}{4}$$

Thus

$$x^2 + 4x + 4 = \frac{11}{4}$$

therefore

$$(x + 2)^2 = \frac{11}{4}$$

and

$$x + 2 = \pm \sqrt{\left(\frac{11}{4}\right)}$$

so

$$x = -2 \pm \sqrt{\left(\frac{11}{4}\right)}$$

which give the roots of the equation.

Example 2.26

Solve $ax^2 + bx + c = 0$, where a, b and c are constants, by completing the square.

Divide through by the coefficient of x^2, that is a, to give

$$x^2 + \frac{b}{a}x + \frac{c}{a} = 0$$

Half the coefficient of x is $\frac{1}{2} \times (b/a)$, that is, $(b/2a)$, so the required square is $[x + (b/2a)]^2$. Now

$$\left(x + \frac{b}{2a}\right)^2 = x^2 + \frac{b}{a}x + \frac{b^2}{4a^2}$$

(square the first term x to give x^2, multiply x and $b/2a$ together and double to give bx/a and square $b/2a$ to give $b^2/4a^2$.) We have

$$x^2 + \frac{bx}{a} + \frac{c}{a}$$

The difference is

$$x^2 + \frac{bx}{a} + \frac{b^2}{4a^2} - x^2 - \frac{bx}{a} - \frac{c}{a}$$

that is

$$\frac{b^2}{4a^2} - \frac{c}{a}$$

Add $[(b^2/4a^2) - (c/a)]$ to both sides of the original equation to give

$$x^2 + \frac{bx}{a} + \frac{c}{a} + \frac{b^2}{4a^2} - \frac{c}{a} = \frac{b^2}{4a^2} - \frac{c}{a}$$

thus

$$x^2 + \frac{bx}{a} + \frac{b^2}{4a^2} = \frac{b^2}{4a^2} - \frac{c}{a}$$

and

$$\left(x + \frac{b}{2a}\right)^2 = \frac{b^2}{4a^2} - \frac{c}{a}$$

so that

$$x + \frac{b}{2a} = \pm\sqrt{\left(\frac{b^2}{4a^2} - \frac{c}{a}\right)}$$

and

$$x = -\frac{b}{2a} \pm\sqrt{\left(\frac{b^2}{4a^2} - \frac{c}{a}\right)}$$

which give the required roots. This expression is in fact the *formula* which may be used to solve any quadratic equation. For ease of remembering it is usually written in a slightly different way. Writing

$$\left(\frac{b^2}{4a^2} - \frac{c}{a}\right) \text{ as} \left(\frac{b^2}{4a^2} - \frac{4ac}{4a^2}\right)$$

which is the same thing, we may write

$$\frac{b^2 - 4ac}{4a^2}$$

and since the square root of $4a^2$ is $2a$, then

$$x = -\frac{b}{2a} \pm \frac{1}{2a}\sqrt{(b^2 - 4ac)}$$

$$x = \frac{-b \pm \sqrt{(b^2 - 4ac)}}{2a}$$

Example 2.27

Solve the equation $3x^2 - 5x - 7 = 0$.

In this equation
    coefficient of x^2 (*a* in the formula) is 3
    coefficient of x (*b* in the formula) is $- 5$ (*include sign*)
    constant (*c* in the formula) is $- 7$ (*include sign*)
Thus

$$x = \frac{-(-5) \pm \sqrt{[(-5)^2 - 4 \times 3 \times (-7)]}}{2 \times 3}$$

$$= \frac{5 \pm \sqrt{(25 + 84)}}{6}$$

$$= \frac{5 \pm \sqrt{(109)}}{6}$$

which gives the required roots.

Example 2.28

Solve $2x^2 + 5x - 3 = 0$ using the formula.

In this equation *a* of the formula is 2, *b* of the formula is 5, *c* of the formula is $- 3$. Thus

$$x = \frac{-5 \pm \sqrt{[25 - 4 \times 2 \times (-3)]}}{4}$$

$$= \frac{-5 \pm \sqrt{(25 + 24)}}{4}$$

$$= \frac{-5 \pm \sqrt{49}}{4} \quad \text{that is} \quad \frac{-5 \pm 7}{4}$$

$$= \frac{-5 + 7}{4} \quad \text{and} \quad \frac{-5 - 7}{4}$$

$$= \tfrac{1}{2} \text{ and } - 3$$

From these examples it can be seen that the formula is a most useful method of solving quadratic equations. **It should be committed to memory**.

Exercise 2.4

(1) Solve the following equations for x by factorisation

(a) $x^2 + 5x + 4 = 0$ (d) $x^2 - (x/4) - (1/8) = 0$
(b) $x^2 - 5x - 6 = 0$ (e) $2x^2 + 5x - 3 = 0$
(c) $2x^2 + 2x - 40 = 0$ (f) $81x^2 + 72x + 16 = 0$

(2) Solve the following equations for x by completing the square

(a) $x^2 + 2x - 3 = 0$ (d) $4x^2 - 28x + 49 = 0$
(b) $x^2 - 10x + 10 = 0$ (e) $x^2 - 0.8x + 0.1 = 0$
(c) $x^2 + x + (\frac{1}{8}) = 0$ (f) $4x^2 + 20x + 19 = 0$

(3) Use the formula to solve the following equations for x

(a) $x^2 + 7x - 3 = 0$ (d) $(x^2/4) - (x/8) - (1/16) = 0$
(b) $5x^2 - 3x - 2 = 0$ (e) $2.3x^2 + 4.1x - 7.8 = 0$
(c) $2x^2 - 12x + 3 = 0$ (f) $0.18x^2 - 0.2x - 0.3 = 0$

(4) Find the value of two resistors that have a total resistance of 18.9 Ω when connected in series and 4.1 Ω when connected in parallel.

(5) In a certain series circuit made up of two resistors the p.d. across one resistor is 12 V and the value of the resistance of the other resistor is 10 Ω. Calculate the possible values of current which would cause the total power dissipation in the circuit to be 10 W.

(6) A rectangular field is 10 m longer than it is wide and the distance between opposite corners along a diagonal is 500 m. Calculate the width of the field.

(7) The equation relating the anode current i_a and the anode voltage V_a of a certain diode is $i_a = 0.004 V_a + 0.0009 V_a^2$. Calculate the anode voltage for an anode current of 10 mA.

(8) Find a positive number such that when it is squared it is greater by 8 than when it is multiplied by four.

(9) Solve the equation

$$\frac{x - 3}{x + 5} = \frac{2x + 2}{x - 1}$$

(10) The equation relating distance s metres travelled by a body in t seconds is $s = 8t + 6t^2$. Calculate the time taken to travel 100 metres.

(11) Two resistors connected in series have a total resistance of 25 Ω and absorb 4 W when the voltage across one resistor is 2 V less than the voltage across the other resistor. Find the total supply voltage under these conditions.

(12) The equation relating the current I through an active component with the p.d. across it, V, is $I = 0.3V^2 - 0.1V$. Find the p.d. across the component when the current is 200 mA.

(13) Find the radius of a cylinder having a total surface area (including both ends) of 100 cm^2 if the cylinder height is 10 cm.

(14) A 20 mH choke of negligible resistance is connected in series with a 0.1 μF capacitor. Find the frequency at which the impedance of the circuit is 30 Ω and inductive.

(15) When the numerical value of the area of a certain circle is added to the numerical value of its circumference the result is 100. Find the circle radius.

3 Graphs

The basic theory of graphs as a method of showing pictorially how the value of one variable quantity varies with the value of another was considered in *First-year Technician Mathematics*. In that book graphs of simple linear equations were plotted and a graphical method of solution of simultaneous linear equations was demonstrated. In this chapter we shall be extending the method to solve quadratic equations and simultaneous equations, both quadratic and linear. In addition we shall be examining the further use of the straight-line graph as a means of solving problems involving non-linear equations.

GRAPHICAL SOLUTION OF QUADRATIC EQUATIONS

By plotting the graph of a quadratic equation of the form $y = ax^2 + bx + c$, in which the value of the *dependent variable y* is determined by the value of the *independent variable x* and the value of the constants a, b and c, we can find the value or values of x that make y have a particular value. We can then use this to solve quadratic equations. For example, if we plot the graph of $y = 2x^2 + 7x + 6$ we can find from the graph the values of x which make $y = 0$ and thus solve the equation $2x^2 + 7x + 6 = 0$. Or, as another example, using the graph we can find the values of x which make $y = 3$ and thus solve the equation

$$2x^2 + 7x + 6 = 3$$

which is

$$2x^2 + 7x + 3 = 0 \quad \text{subtracting 3 from both sides}$$

The method is demonstrated in the following examples.

Example 3.1

Plot the graph of $y = 2x^2 + 7x + 6$ for values of x between -5 and 2 and thus solve the equations (a) $2x^2 + 7x + 6 = 0$ and (b) $2x^2 + 7x + 3 = 0$

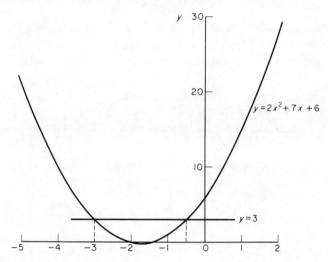

Figure 3.1

The graph is shown in figure 3.1 and the table 3.1 lists the values. Note that x is increased in increments of $\frac{1}{2}$ to give a total of 15 points on the graph so that a good curve may be obtained.

Table 3.1

x	-5	$-9/2$	-4	$-7/2$	-3	$-5/2$	-2	$-3/2$	-1	$-1/2$	0	1/2	1	$-3/2$	2
$2x^2$	50	81/2	32	49/2	18	25/2	8	9/2	2	1/2	0	1/2	2	9/2	8
$7x$	-35	$-63/2$	-28	$-49/2$	-21	$-35/2$	-14	$-21/2$	-7	$-7/2$	0	7/2	7	21/2	14
6	6	6	6	6	6	6	6	6	6	6	6	6	6	6	6
y	21	15	10	6	3	1	0	0	1	3	6	10	15	21	28

   (a) From the table of values and from the graph we see that $y = 0$ when $x = -2$ and when $x = -3/2$ (shown on the graph where the x axis, at which $y = 0$, cuts the curve). The roots of the equation $2x^2 + 7x + 6 = 0$ are therefore -2 and $-3/2$.

   (b) To obtain the roots of $2x^2 + 7x + 3 = 0$ from the graph of $y = 2x^2 + 7x + 6$ take the equation $2x^2 + 7x + 3 = 0$ and add 3 to both sides to give $2x^2 + 7x + 6 = 3$. Thus when

$$2x^2 + 7x + 6 = 3$$

$$2x^2 + 7x + 3 = 0$$

and

$$2x^2 + 7x + 6 = 3$$

when y in the equation $y = 2x^2 + 7x + 6$ is equal to 3. Draw the line $y = 3$ on the same axis as the plotted graph (figure 3.1) and read off x at the points of intersection. These values are $-\frac{1}{2}$ and -3 (also shown in the table of values). Thus when $x = -\frac{1}{2}$ and -3

$$y = 2x^2 + 7x + 6 = 3$$

and

$$2x^2 + 7x + 3 = 0$$

Thus the roots of $2x^2 + 7x + 3 = 0$ are $x = -\frac{1}{2}$ and $x = -3$.

Example 3.2
Plot the graph of $y = x^2 - 7x + 12$ between $x = 0$ and $x = 6$. (a) Use the graph to find the roots of the equations (a) $x^2 - 7x + 8 = 0$ (b) $x^2 - 7x + 9 = 0$. (b) What is the minimum value of y?

The table of values is shown in table 3.2.

Table 3.2

x	0	1	2	3	4	5	6
x^2	0	1	4	9	16	25	36
$-7x$	0	-7	-14	-21	-28	-35	-42
$+12$	12	12	12	12	12	12	12
y	12	6	2	0	0	2	6

The graph is shown in figure 3.2.

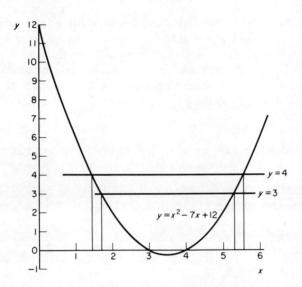

Figure 3.2

(a)   (i) Now if $x^2 - 7x + 8 = 0$ then, by adding 4 to both sides

$$x^2 - 7x + 12 = 4$$

that is, the equation $x^2 - 7x + 8 = 0$ is satisfied when $x^2 - 7x + 12 = 4$ or, in other words, when the graph of $y = x^2 - 7x + 12$ cuts the graph of $y = 4$. As shown in the figure the values of x when this occurs are 1.44 and 5.56

(ii) Similarly the roots of $x^2 - 7x + 9 = 0$ are found by noting the value of x when the curve $y = x^2 - 7x + 12$ cuts $y = 3$. These roots are 1.7 and 5.3.

(b) By examination the minimum value of y, that is, the value of y at the lowest point on the graph, is $-\frac{1}{4}$.

From the above examples we can see that quadratic equations of the form $ax^2 + bx + c = 0$, where a, b and c are constants, may be solved either by plotting the graph of $y = ax^2 + bx + c$ and noting the values of x at the points of intersection of the graph with the x axis (that is, where $y = 0$) or, if another graph is already available of the function $y = ax^2 + bx + d$, where a and b are the constants above and d is a separate constant, by noting the points of intersection of the graph of $y = ax^2 + bx + d$ and the graph of $y = d - c$. For when

$$ax^2 + bx + d = d - c$$

then

$$ax^2 + bx + c = 0$$

which is the equation to be solved. In these two examples we have used a graph of a function such as $y = ax^2 + bx + d$ to solve quadratic equations of the form $ax^2 + bx + c = 0$; that is, the coefficients of x^2 and x are the same in both graph function and equation. We shall now use the same approach to the solution of quadratic equations in which the coefficient of x is different from the coefficient of x in the function of which the graph is available.

Example 3.3

Use the graph of the function $y = x^2 - 7x + 12$ to solve the quadratic equation $x^2 - 8x + 12 = 0$.

Adjust the equation to have one side the same as the function $y = x^2 - 7x + 12$. This can be done by adding x to both sides of $x^2 - 8x + 12 = 0$ to give

$$x^2 - 8x + x + 12 = x$$

that is

$$x^2 - 7x + 12 = x$$

So that for values of x which satisfy the equation $x^2 - 7x + 12 = x$ the equation $x^2 - 8x + 12 = 0$ is also true. These values may be found from the points of intersection of the graphs $y = x^2 - 7x + 12$ and $y = x$. The table of values for the first graph was given in the previous example. Both graphs are shown in figure 3.3.

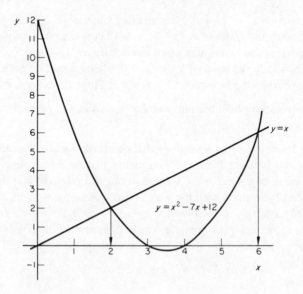

Figure 3.3

At the points of intersection, $x = 2$ and 6. Thus these values are the roots of the equation $x^2 - 8x + 12 = 0$. Clearly, these values could have been obtained from the table of values since these points of intersection coincide with points already plotted. However, this is unusual and in cases where the second graph has a function more involved than $y = x$, solution by inspection of the table of values is more difficult. This is shown in the next example.

Example 3.4

Plot the graph of $y = x^2 - 5x + 4$ for values of x from 0 to 6. Use the graph to solve the equation $x^2 - 6x + 6 = 0$.

The table of values for $y = x^2 - 5x + 4$ is given in table 3.3. The graph is shown in figure 3.4.

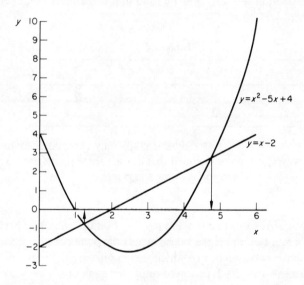

Figure 3.4

Table 3.3

x	0	1	2	3	4	5	6
x^2	0	1	4	9	16	25	36
$-5x$	0	-5	-10	-15	-20	-25	-30
4	4	4	4	4	4	4	4
y	4	0	-2	-2	0	4	10

To solve the equation $x^2 - 6x + 6 = 0$ adjust the equation to read the same on one side as the given function of x given above, that is, $x^2 - 5x + 4$. This is done by adding $(x - 2)$ to both sides of the equation $x^2 - 6x + 6 = 0$ to give

$$x^2 - 6x + 6 + x - 2 = x - 2$$

that is

$$x^2 - 5x + 4 = x - 2$$

The values of x for which this new equation is true are thus the roots of the equation $x^2 - 6x + 6 = 0$. These values are obtained from the points of inter-section of the graph $y = x^2 - 5x + 4$ and $y = x - 2$ since at these points $x^2 - 5x + 4 = y = x - 2$.

Table 3.4 gives values of $y = x - 2$ and the graph is drawn in figure 3.4. At the points of intersection, $x = 1.27$ and 4.73 and thus these are the required roots.

Table 3.4

x	0	1	2	3	4	5	6
-2	-2	-2	-2	-2	-2	-2	-2
y	-2	-1	0	1	2	3	4

Note that since $y = x - 2$ is a straight-line graph only two points are required to locate it. However, it is recommended that at least three points are taken, to reduce the possibility of error in drawing the graph.

Example 3.5

Plot the graphs of $y = x^2 - 3x - 4$ and $y = x - 2$ on the same axes between $x = -3$ and $x = 5$. Determine the values of x at the points of intersection and give the quadratic equation in x of which these values are the roots. How could the equation $x^2 - 4x - 3 = 0$ be solved using the graph of $y = x^2 - 3x - 4$?

Table 3.5 gives values of both functions, the graphs being shown in figure 3.5.

Table 3.5

x	-3	-2	-1	0	1	2	3	4	5
x^2	9	4	1	0	1	4	9	16	25
$-3x$	9	6	3	0	-3	-6	-9	-12	-15
-4	-4	-4	-4	-4	-4	-4	-4	-4	-4
y	14	6	0	-4	-6	-6	-4	0	6

Table 3.5 cont.

x	−3	−2	−1	0	1	2	3	4	5
−2	−2	−2	−2	−2	−2	−2	−2	−2	−2
y	−5	−4	−3	−2	−1	0	1	2	3

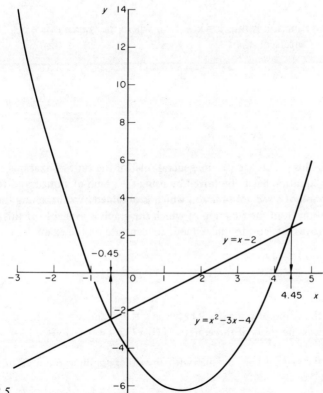

Figure 3.5

From the graphs the values of x at the points of intersection are -0.45 and 4.45. At these points

$$x^2 - 3x - 4 = x - 2$$

that is

$$x^2 - 4x - 2 = 0$$

and this is the quadratic equation in x of which -0.45 and 4.45 are the roots.

To solve the equation $x^2 - 4x - 3 = 0$ using the graph of $y = x^2 - 3x - 4$ determine what function of x the expression $(x^2 - 3x - 4)$ must equal for $x^2 - 4x - 3 = 0$ to be true; the graph of this function is then drawn and the values of x at the points of intersection are the roots of the given equation. The function of x may be obtained by subtracting $x^2 - 4x - 3$ from $x^2 - 3x - 4$ to give

$$(x^2 - 3x - 4) - (x^2 - 4x - 3) = x - 1$$

The required function is thus $y = x - 1$ for where the graphs of $y = x^2 - 3x - 4$ and $y = x - 1$ intersect, then

$$x^2 - 3x - 4 = x - 1$$

and

$$x - 4x - 3 = 0$$

as required. This example gives the general rule for the type of example in which a quadratic equation has to be solved by using the graph of a quadratic function. Draw the graph of a second function which is obtained by subtracting the terms of the equation from the function of which the graph is available. If this general approach proves difficult to understand, concentrate on the examples given here.

Example 3.6

(a) Plot the graph of $y = x^2 - 9$ between $x = -4$ and $x = +4$ and show how the graph gives the roots of the equations (i) $x^2 - 9 = 0$ (ii) $x^2 - 13 = 0$.

(b) Use the given graph together with any other graph, as required, to solve $x^2 - x - 8 = 0$.

(a) Table 3.6 gives values of $y = x^2 - 9$.

Table 3.6

x	-4	-3	-2	-1	0	1	2	3	4
x^2	16	9	4	1	0	1	4	9	16
-9	-9	-9	-9	-9	-9	-9	-9	-9	-9
y	7	0	-5	-8	-9	-8	-5	0	7

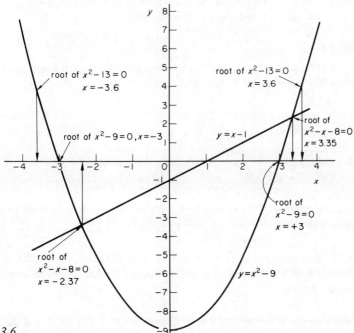

Figure 3.6

The graph is shown in figure 3.6. $x^2 - 9 = 0$ where the graph of $y = x^2 - 9$ cuts the x axis (at which $y = 0$). At these points $x = 3$ and -3. Similarly $x^2 - 13 = 0$ where $x^2 - 9 = 4$ so that by reading off values of x when $y = 4$ on the graph the roots of $x^2 - 13 = 0$ will be obtained. As shown these are $x = 3.6$ and $x = -3.6$.

(b) To solve $x^2 - x - 8$ plot the graph of

$$y = (x^2 - 9) - (x^2 - x - 8)$$

$$= x - 1$$

and note the values of x at the intersection of $y = x^2 - 9$ and $y = x - 1$, since here $x^2 - 9 = x - 1$ and $x^2 - x - 8 = 0$. Since $y = x - 1$ is linear only three points are required (table 3.7).

Table 3.7

x		-3		0		3
-1		-1		-1		-1
y		-4		-1		2

The graph is shown in figure 3.6. At the points of intersection, $x = -2.87$ and 3.35 which are the required roots of the equation $x^2 - x - 8 = 0$.

Exercise 3.1

(1) Plot the graph of $y = x^2 - 8x + 15$ between $x = 0$ and $x = 5$ and use it to solve the equation $x^2 - 7x + 12 = 0$.

(2) Plot the graph of $2y = 2x^2 - 3x - 5$ between $x = -1$ and $x = 2$ and use it to solve the equation $x^2 - \frac{1}{2}x - \frac{1}{2} = 0$.

(3) Solve the equation $x^2 - x - 2 = 0$ using the graph of $y = x^2 - 6$ plotted between $x = -2$ and $x = 3$.

(4) Using the graph of $y = 2x^2 - 12x + 14$ plotted for values of x between 0 and 5 solve the equations (a) $2x^2 - 12x + 14 = 0$ and (b) $2x^2 - 11x + 12 = 0$.

(5) Plot the graphs of $y = 2x - 5$ and $y = x^2 + 8x$ between $x = 0$ and $x = -6$ and note the values of x at the points of intersection. Of what quadratic equation are these values the roots?

(6) Plot the graph of $y = x^2 + \frac{3}{4}x + \frac{1}{8}$ between $x = -2$ and $x = 2$ and use the graph to solve the equations $x^2 + \frac{1}{4}x - \frac{1}{8} = 0$ and $x^2 + \frac{3}{4}x + \frac{1}{8} = 0$.

(7) Solve $x^2 - x - 2 = 0$ and $x^2 - 4x - 8 = 0$ using the graph of $y = x^2 - 4x - 8$ between $x = -2$ and $x = 3$.

(8) Plot the graph of $y = x^2 + 4x - 68$ for values of x between -10 and 10 and use this graph to solve the equation $x^2 + 2x - 63 = 0$.

(9) Solve the simultaneous equations $y = x^2 - 3x - 63$ and $y = 2x + 21$ by drawing the graphs of these functions between $x = -8$ and $x = 13$. Of what quadratic equation are these values of x the roots?

(10) Show the roots of $x^2 - 9 = 0$ on the graph of $y = x^2 + x - 4$ drawn between $x = -4$ and $x = 4$.

USE OF STRAIGHT-LINE GRAPHS FOR NON-LINEAR RELATIONSHIPS

In *First-year technician Mathematics* the general equation of a straight-line graph was shown to be of the form

$$y = ax + b$$

where y is a variable having a value determined by the values of variable x and constants a and b. Plotting y vertically as ordinate and x horizontally as abscissa the constant a is the *gradient* or slope of the straight-line graph. The constant b is the value of y when $x = 0$, that is, it is the value of y on the y axis where the graph cuts the axis*; b is thus called the *intercept* of the graph on the y axis. The equation $y = ax + b$, having no power of x other than unity, is called a linear equation and we say that the variable y has a *linear* relationship with the variable x.

Straight-line graphs are by far the easiest of all graphs to draw and to recognise in problems where the relationship between two variables is not known and an attempt is being made to determine it by drawing a graph. This point may be illustrated by considering the graphs shown so far in this chapter, all of which were of quadratic relationships (involving x^2). Many points were needed to obtain a good curve and the curve was not particularly easy to draw. Also if we were attempting to deduce a relationship between variables from such a graph shape we could not be absolutely sure that the relationship was indeed quadratic without further tests. On the other hand a straight-line graph of a relationship known to be linear requires only two or, at the most, three points to establish it and in attempting to deduce the form of an unknown relationship the linear nature of the graph is easily recognisable. These characteristics of straight-line graphs enable them to be used to good effect in the solution of problems involving relationships which would not give straight-line graphs if one variable were plotted directly against another. The following examples should be studied with care.

Example 3.7

The power absorbed by a certain resistor for various values of current is as follows.

Power (watts)	5	20	45	80	125	180
Current (amperes)	1	2	3	4	5	6

Plot the graph of power against the square of the current and hence determine the equation relating power and current for this resistor.

The required values for the graph are as follows.

Power	5	20	45	80	125	180
Current	1	2	3	4	5	6
(Current)^2	1	4	9	16	25	36

---

*This is true provided, of course, that the scales are such that $x = 0$ at the y axis. Occasionally this is not so, as was explained in *First-year Technician Mathematics*.

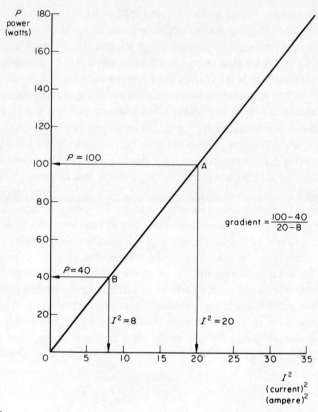

Figure 3.7

The graph is shown in figure 3.7. As shown it is a straight-line graph, the intercept on the power axis being zero. Denoting power by P and current by I the equation is therefore

$$P = \text{constant} \times I^2$$

since plotting P against I^2 gives a straight line. The constant is the gradient of the graph and, taking the two points A and B (points that are *not* given in the table of values)

at A

$$P = 100 \qquad\qquad I^2 = 20$$

At B

$$P = 40 \qquad\qquad I^2 = 8$$

The gradient is thus $(100 - 40)/(20 - 80)$ that is, 5 and the equation is $P = 5I^2$. (In this particular example 5 is actually the value of the resistance in ohms since, from electrical theory, power = resistance $\times$ current^2.)

Note that in this particular example the values given yield a graph which is a perfectly straight line. Often, experimental error gives results that produce a variation in the position of points such that a perfectly straight line does not pass through all of them. Provided the variation is not excessive a linear relationship may be inferred and the equation deduced from the straight line drawn so as to pass through as many of the points as possible with as many of the remaining points above the line as below it, as far as this can be arranged. This is shown in the next example.

Example 3.8

The opposition to electric current, Z, measured in ohms, at different values of frequency, f, measured in hertz, of the current is as follows.

Z (ohms)	400	205	130	104	75	67
f (hertz)	25	50	75	100	125	150

By plotting a straight-line graph show that the equation connecting Z and f is $Z = k/f$ where k is a constant. Find the value of k.

To show that $Z = k/f$, a graph plotting values of Z against values of $1/f$ is drawn. If this graph is a straight line

$$Z \propto \frac{1}{f}$$

(that is, Z is proportional to $1/f$) and

$$Z = \frac{k}{f} \quad \text{where } k \text{ is a constant}$$

An extended table of values is required as follows.

Z	400	205	130	104	75	67
f	25	50	75	100	125	150
$1/f$	0.04	0.02	0.013	0.01	0.008	0.0067

For convenience the last row may be expressed as a number and multiplier, a convenient multiplier being 10^{-2}, to give

$1/f$	4	2	1.3	1.0	0.8	0.67	$(\times 10^{-2})$

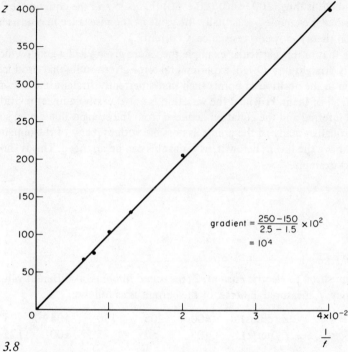

Figure 3.8

The graph is shown in figure 3.8. The line is drawn such that any points that do not lie on it are, as far as possible, equally displaced from it. The gradient, using points (not in the table of values) which have the co-ordinates $Z = 250$, $1/f = 250 \times 10^{-2}$ and $Z = 150$, $1/f = 1.5 \times 10^{-2}$, is given by

$$\frac{250 - 150}{(2.5 - 1.5) \times 10^{-2}}$$

which equals 10^4. The equation is thus

$$Z = \frac{10^4}{f}$$

Example 3.9

The total resistance, R, of two pieces of wire connected in series is believed to be related to the diameter, d, of one of the pieces of wire, when all other dimensions remain constant, by an equation of the form

$$R = \frac{a}{d^2} + b$$

where a and b are constants. Observed values of R and d are as follows.

R (ohms)	149	76	60	58	54
d (cm)	0.1	0.2	0.3	0.4	0.5

Using a straight-line graph find whether or not the form of the equation is correct. If it is, find the values of a and b.

Plotting R against d would give a non-linear curve and would not help to verify that the given equation form is valid. However, if R is plotted against $(1/d^2)$ and the equation is correct, a straight line should result with gradient a and intercept (on the R axis) b. This may be seen by comparing the equation

$$R = a\left(\frac{1}{d^2}\right) + b$$

with the standard $y = mx + c$ in which m is the gradient and c the intercept on the y axis.

An extended table of values is given in table 3.8.

Table 3.8

R	149	76	60	58	54
d	0.1	0.2	0.3	0.4	0.5
d^2	0.01	0.04	0.09	0.16	0.25
$1/d^2$	100	25	11.1	6.25	4

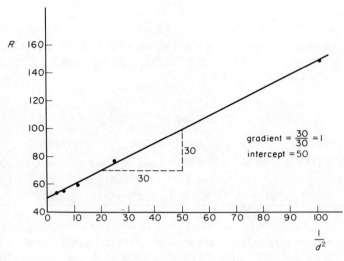

Figure 3.9

The graph is shown in figure 3.9. As shown the gradient is 1 and the intercept when $1/d^2 = 0$ is 50. The equation is therefore

$$R = \frac{1}{d^2} + 50$$

Examples 3.7, 3.8 and 3.9 illustrate the general approach to the use of straight-line graphs in validating non-linear equations. The method is not to plot one variable against another, which would yield a non-linear graph, but to plot *a function* of one variable against the other chosen so as to obtain a straight-line graph. The function chosen — the reciprocal of the variable, as in example 3.8; the square of the variable, as in example 3.7; or the reciprocal of the square of the variable, as in the last example — is determined by what is necessary to make a linear equation from the non-linear equation that is to be validated.

For example, in example 3.7, the equation was

$$P = kI^2$$

Writing I^2 as x we obtain $P = kx$ which if P is plotted against x gives a straight-line graph passing through the origin and having gradient k. Thus, plot P against x, that is, P against I^2, and if the graph is linear, the equation $P = kI^2$ is true.

In example 3.9, the equation was

$$Z = \frac{k}{f}$$

Writing $1/f$ as x we obtain $Z = kx$ which if Z is plotted against x gives a straight-line graph passing through the origin and having gradient k. Thus, plot Z against x, that is, Z against $1/f$, and if the graph is linear, the equation $Z = k/f$ is valid.

In example 3.9, the equation was

$$R = \frac{a}{d^2} + b$$

and writing $1/d^2$ as x we obtain $R = ax + b$ which, if R is plotted against x, gives a straight-line graph cutting the R axis at b and having a gradient a. Thus, plot R against x, that is, R against $1/d^2$, and if the graph is linear the equation $R = (a/d^2) + b$ is true.

Other possible non-linear equations that can be tested using straight-line graphs include $y = a\sqrt{x} + b$ in which a, b are constants (a straight-line graph is obtained if y is plotted against $\sqrt{x}$) and $y = Ax^n$, in which A and n are constants. The next section is devoted to this equation.

Laws of the form $y = Ax^n$

The shape of the curve produced by plotting y against x when y and x are related by an equation of the form $y = Ax^n$ is determined largely by the values of A and n, particularly the latter. For example, if $n = 1$ the graph is a straight-line of gradient A which passes through the origin. For other values the curve is non-linear and if n is known a straight line may be obtained by plotting y against x^n. This was done in example 3.7 in which $n = 2$ (since $P = kI^2$); example 3.8 in which $n = -1$ (since $Z = k/f$) and 3.9 in which $n = -2$ (since $R = (a/d^2) + b$). However, when n is to be determined by the use of a straight-line graph, the value of x cannot be calculated and plotting y against x is not possible. In these cases the usual method is to take logarithms of both sides, for if $y = Ax^n$

$$\log y = \log (Ax^n)$$

$$= \log A + n \log x$$

(Logarithms of numbers to be multiplied are added, and the logarithm of a number raised to a power is given by the power multiplied by the logarithm of that number.) Thus if $\log y$ is plotted against $\log x$ we have a linear relationship in which n is the gradient and $\log A$ the intercept on the $\log y$ axis. To clarify this consider the general form of a linear equation

dependent variable = gradient $\times$ independent variable + intercept

usually written

$$y = mx + c$$

and in this case

$$\log y = n (\log x) + \log A$$

that is, $\log y$ taking the place of y, n taking the place of m, $\log x$ taking the place of c.

Example 3.10

The anode current I_A and the anode voltage V_A of a diode valve are related by the equation

$$I_A = kV_A^n$$

where I_A is measured in milliamperes and V_A in volts over the range of measurements given below.

I_A (mA)	1	2	3	4	5
V_A (V)	4.67	7.41	9.55	11.75	13.49

Determine values of k and n by using a suitable straight-line graph.

Taking logarithms of both sides of the equation

$$\log I_A = n \log V_A + \log k$$

The graph of $\log I_A$ against $\log V_A$ using the observed values should be a straight line of gradient n and intercept $\log k$ if this equation is valid. An extended table of values follows.

I_A (mA)	1	2	3	4	5
V_A (V)	4.67	7.41	9.55	11.75	13.49
$\log I_A$	0	0.301	0.4771	0.6021	0.699
$\log V_A$	0.6693	0.8698	0.98	1.07	1.13

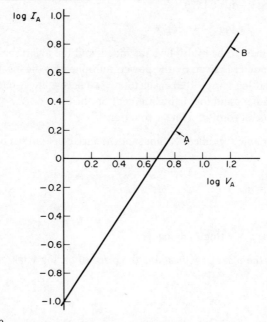

Figure 3.10

The graph is shown in figure 3.10. Note that the $\log I_A$ axis has to be extended in the negative direction in order to be able to read off the intercept directly. (The meaning of a negative intercept in this type of problem is considered later.)

At point A

$$\log I_A = 0.2 \qquad \text{and } \log V_A = 0.8$$

At point B

$$\log I_A = 0.8 \qquad \text{and } \log V_A = 1.2$$

The gradient is thus

$$\frac{0.8 - 0.2}{1.2 - 0.8} \qquad \text{that is,} \qquad \frac{0.6}{0.4}$$

which equals $3/2$. The value of n is therefore $3/2$.

The intercept of the graph on the $\log I_A$ axis is -1 which tells us that $\log k = -1$. Remembering that a logarithm consists of two parts — the mantissa, which is always positive and follows the decimal point, and a characteristic, which may be positive or negative and which precedes the decimal point — then if we have $\log k = -1$, -1 is the characteristic (usually written as $\bar{1}$ and called 'bar one') and .0000 is the mantissa. Thus $\log k = \bar{1}.0000$ and $k = 0.1$ Therefore the equation is

$$I_A = 0.1 \, V_A^{3/2}$$

It is worth considering at this point how k could be determined had the intercept been, say, -0.8. Now as stated above a logarithm has two parts, one positive or negative, the other always positive. Consequently -0.8 is inadmissible and we must write it as $-1 + 0.2$, that is, $\bar{1}.2$ which would give $\log k = \bar{1}.2$ and $k = 0.1585$ from antilogarithm tables.

Example 3.12

The luminous intensity I (candela) of a lamp is believed to be related to the lamp voltage V by an equation of the form $I = kV^n$ where k and n are constants. Values of I and V are as follows.

I (cd)	6.48	20.48	50	103.7
V (V)	60	80	100	120

Find the values of k and n.

If $I = kV^n$ then $\log I = n \log V + \log k$ so that plotting $\log I$ against $\log V$ would produce a straight line of gradient n and intercept on the $\log I$ axis (where $\log V$ is zero) of $\log k$. The table of values is

$\log I$	0.8116	1.3114	1.6990	2.0157
$\log V$	1.7782	1.9031	2	2.0792

To achieve good spacing between points and obtain the best linear graph possible it is necessary with the values in this example to begin both scales at values greater than zero, as shown in figure 3.11. As was stated in earlier work

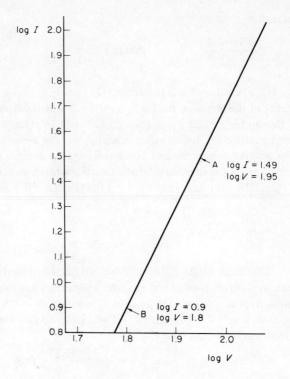

Figure 3.11

and again at the beginning of this section on line graphs, the constant c in the general equation $y = mx + c$ is the intercept on the y axis *provided that x is 0 at the y axis*. In this example $\log V$ is not 0 at the $\log I$ axis and the value of the constant, $\log k$, cannot be obtained by examination of the figure.

Points A and B on the figure give $\log I = 1.49$ and 0.90, $\log V$ having the values 1.95 and 1.80 respectively. We may thus find the values of n and $\log k$ by inserting these values into the equation

$$\log I = n \log V + \log k$$

to give two simultaneous equations as shown below. (The method employed here is very useful and should be employed in cases where the axes do not intersect at the point $(0, 0)$. Great care must always be taken when reading intercepts that the abscissa variable is in fact zero at the point of intersection. If it is not the simultaneous equation method must be used.)

The two equations are

$$1.49 = 1.95n + \log k \tag{1}$$

$$0.9 = 1.8n + \log k \tag{2}$$

Subtracting equation 2 from equation 1

$$0.59 = 0.15n$$

and

$$n = \frac{0.59}{0.15}$$

$$= 3.933$$

Substituting in equation 1 for n

$$1.49 = 1.95 \times 3.93 + \log k$$

that is

$$1.49 = 7.670 + \log k$$

and

$$\log k = 1.49 - 7.670$$

$$= -6.18$$

This is a wholly negative number and is inadmissible as a logarithm, which must have a wholly positive section (the mantissa) after the decimal point. To write -6.1799 in logarithmic form obtain the characteristic by changing -6 to -7 (bar 7) and then determine what positive number must be added to -7 to give -6.18, that is, the mantissa is $7 - 6.18 = 0.82$. Thus -6.18 may be written as $-7 + 0.82$ that is, $\bar{7}.82$ and

$$\log k = \bar{7}.82$$

so that

$$k = 6.609 \times 10^{-7}$$

The equation is thus $I = 6.609 \times 10^{-7} \times V^{3.933}$

USE OF LOGARITHMIC GRAPH PAPER

In examples 3.10 and 3.11, in which straight-line graphs were used to find the constant A and index n in equations of the form $y = Ax^n$, a table of values containing the logarithms of x and y for various values of x and y was first prepared. To avoid the use of logarithms and to reduce the time and effort needed in solution of such problems, logarithmic graph paper may be employed.

All the graphs so far considered both in this book, and in *First-year Technician Mathematics*, used scales such that the distance along the axis between any two points is directly proportional to the size of the variable quantity being represented.

Thus if, for example, 1 cm represents 10 V, then 2 cm represents 20 V, 3 cm represents 30 V, and so on. Such scales are called *linear* and graph paper with equally spaced graduations to accommodate these scales is called linear graph paper. A logarithmic scale is one such that the distance along the axis to the point representing a particular value of a variable quantity is proportional to the logarithm of the value and not the value itself. Logarithmic graph paper, one type of which is illustrated in figure 3.12 has its graduations marked so that the use of a logarithmic scale is possible.

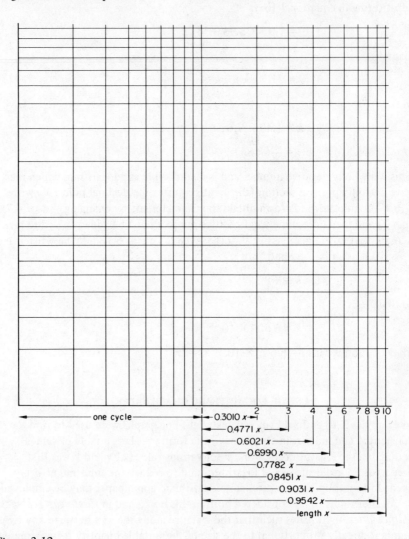

Figure 3.12

Logarithmic graph paper is divided into equal scale lengths called *cycles,* each cycle representing 10 X the value of that represented by the cycle on the left (or below) and 0.1 X the value of that represented by the cycle on the right (or above). The paper shown in the figure is 2 X 2-cycle paper. Each cycle is subdivided by graduations placed at intervals corresponding to the logarithm of the number of the graduation. Thus graduation 1 lies at the beginning of the cycle (log 1 = 0), graduation 2 lies a distance of 0.3010 of the cycle length (shown as x in the figure) along the axis from the beginning of the cycle and corresponding distances from the beginning to graduations 3, 4, 5, etc., equal to $0.4771x$, $0.6021x$, $0.6990x$, and so on, until graduation 10 lies at the end of the cycle (log 10 = 1). The value of the variable represented is always equal to the number of the graduation multiplied by a power of ten, the power being chosen so as to accommodate whatever range of values is required. The power of ten in the multiplier rises by 1 as one progresses to the right along the horizontal axis or upwards along the vertical axis.

To illustrate this consider using logarithmic graph paper to plot the following values.

x	3	8	12	20	30
y	19	41	81	112	150

The values of x lie between 1 and 100, two lying between 1 (10^0) and 10 (10^1) and three lying between 10 and 100 (10^2). Two cycles would be used for x, the first containing values from 1 to 10, the second containing values from 10 to 100. The graduations in the first cycle represent values of x

1   2   3   4   5   6   7   8   9

the multiplier being 10^0, that is, 1; the graduations in the second cycle represent values of x

10   20   30   40   50   60   70   80   90

the multiplier for this cycle being 10^1, that is, 10. The values of y lie between 10 and 1000, three lying between 10 (10^1) and 100 (10^2) and two lying between 100 and 1000 (10^3). There are no values in the range 1 to 10. Again two cycles are needed, the first containing values from 10 to 100, the second containing values from 100 to 1000. The graduations in the first cycle represent values of y

10   20   30   40   50   60   70   80   90

the multiplier being 10^1; the graduations in the second cycle represent values of y

100   200   300   400   500   600   700   800   900

the multiplier for this cycle being 10^2. (Note that only part of the second cycle would be needed but 2 X 2-cycle paper would be used since part cycles are not usually available.)

As indicated, each cycle represents a *decade* of values, that is, a range between integral powers of ten. The cycles would similarly be used for the following fractional ranges and integral multipliers of them

0.001	to	0.01
0.01	to	0.1
0.1	to	1

One of the advantages of using logarithmic graph paper is that where a linear relationship exists between the logarithms of two variables — that is, one which produces a straight-line graph when the logarithm of one variable is plotted against the logarithm of the other — then plotting the *values* of the variables (*not* their logarithms) on to logarithmic graph paper will produce a straight line. This is because the placing of each graph point is determined horizontally by the logarithm of the x-variable and vertically by the logarithm of the y-variable, since the graduations used for marking values are displaced in space according to a logarithmic law. In effect the graph 'takes logarithms' for us. A similar situation in which lengths are marked according to logarithmic values exists with the slide rule, and here lengths of rule may be added or subtracted to give multiplication or division of the numbers marked on the rule (see *First-year Technician Mathematics*, chapter 1). Since values of the variables may be plotted directly on to the graph, the taking of logarithms from tables is rendered unnecessary.

Before looking at an example involving the use of logarithmic graph paper it is advisable to examine the method of locating points on the graph. When plotting graphs it is often necessary to estimate the position of a particular value of one of the variables when that value lies between the given graduations. For example, suppose one wishes to locate $x = 35$ and the graduations are marked $x = 30$ and $x = 40$. Clearly $x = 35$ will lie halfway between the two values, if a linear scale is used, or, more specifically since the distance between graduations represents a change in value of x of 10, then half the distance represents 5 and this added to the length representing $x = 30$ will give $x = 35$. With logarithmic graph paper the lengths along the axes are proportional to the logarithms of the values of the variables rather than the values themselves and a linear estimate of the position of a value between graduations would be incorrect.

For example consider an axis of length 3 cycles representing 1 to 10, 10 to 100 and 100 to 10000. The axis begins at the value 1, the logarithm of the value 1 being 0. The first cycle represents 10, the logarithm of which is 1; the second cycle terminates at value 100, the logarithm of which is 2; the third cycle terminates at value 1000, the logarithm being 3 (see figure 3.13). If x is the actual length taken to represent one cycle then the axis length from value 1 is x at the end of the first cycle, $2x$ at the end of the second cycle and $3x$ at the end of the third. The length of axis between value 1, the beginning, and any particular graduation is proportional to the logarithm of the value represented by the graduation. Thus

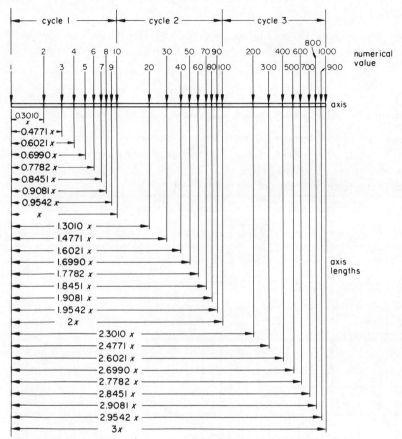

Figure 3.13

the axis length between value 1 and value 5 is $0.6990x$, between value 1 and value 1.5 is $1.6990x$ and between value 1 and value 2.5 is $2.6990x$. The remaining lengths are as shown in the figure. If the position of a value lying between graduations is required then the axis length may be determined using this fact. Thus the axis length to, say, value 8.5 will be $0.9294x$ (since log 8.5 = 0.9294); and clearly, whereas on a linear scale 8.5 lies midway between 8 and 9, here 8.5 lies at length $0.9294x$ which is *not* midway between 8 lying at $0.9081x$ and 9 lying at $0.9542x$. Similarly 15 which is midway between 10 and 20 will lie at a point $1.1761x$ from value 1 (log 15 = 1.1761) which is not midway between x where 10 lies and $1.3010x$ where 20 lies.

As far as possible it is advisable to use only marked graduations and avoid estimating of position between graduations, but if it is unavoidable, the direct proportion between length and logarithm of value represented *must* be borne in mind, otherwise false assumptions may be made and false conclusions drawn.

Example 3.13

Two variable quantities p and v are believed to be related by an equation of the form $p = Av^n$ where A and n are constants. Observed values of p and v are as follows.

p	34	58	96	156	182
v	2.89	8.41	23.04	60.84	82.81

Determine whether the equation is valid and if it is find the values of A and n.

The values of p lie between 10 and 1000 and the p axis therefore requires two cycles, values 10 to 100 and 100 to 1000. Similarly the v axis requires two cycles but with values 1 to 10 and 10 to 100. The graph is shown in figure 3.14 and is linear. The equation is thus of the form $p = Av^n$.

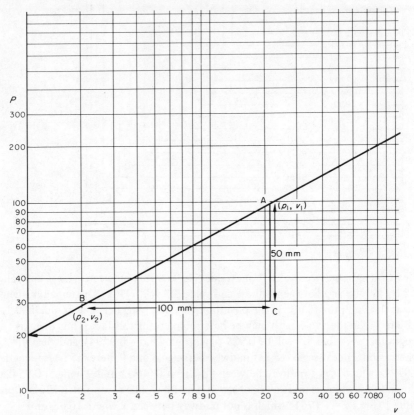

Figure 3.14

Now, if $p = Av^n$ then $\log p = n \log v + \log A$, and for any two points A and B having coordinates p_1, v_1 and p_2, v_2 respectively (that is, p has the value p_1 at A and p_2 at B, v has the value v_1 at A and v_2 at B) then

$$\text{gradient } n = \frac{\log p_1 - \log p_2}{\log v_1 - \log v_2}$$

On our graph the difference between $\log p_1$ and $\log p_2$ is proportional to the length AC since the vertical distance between C and the v axis is proportional to $\log p_1$ and the vertical distance between B and the v axis is proportional to $\log p_2$. Similarly the difference between $\log v_1$ and $\log v_2$ is proportional to length BC since the horizontal distance between B and the p axis is proportional to $\log v_2$ and between C and the p axis is proportional to $\log v_1$. Thus

$$\text{gradient } n = \frac{\text{length AC}}{\text{length BC}}$$

$$= \frac{50 \text{ mm}}{100 \text{ mm}}$$

$$= 0.5$$

(*Note* When determining gradients in other similar problems the gradient is always the ratio between lengths, and not between values represented by lengths as with linear graph paper.)

From the equation $\log p = n \log v + \log A$ it is clear that when $\log v = 0$, $\log p = \log A$. Now $\log v = 0$ when $v = 1$ and this is the origin on the graph in figure 3.14. Thus where our graph cuts the p axis, $\log p = \log A$ or $p = A$ and since the axis graduations denote values of p, the value of A may be read off directly. As shown, $A = 20$ and the equation is thus $p = 20 \, v^{0.5}$. (*Note* The value of the constant may only be read off directly if the horizontal axis begins at value 1, that is, where the logarithm of the value is zero. This situation is directly comparable to that met in solving an equation of the form $y = mx + c$ on linear graph paper where c can only be read off at the intercept provided that the x axis begins at $x = 0$, for only when $x = 0$ does $y = c$.)

Exercise 3.2

(1) The values of the current I and resistance R in a certain circuit with constant applied voltage V are as follows

I (A)	0.1	0.1	0.5	0.7	0.9
R (Ω)	70	23	14.1	10	7.8

Verify, by drawing a straight-line graph, that $I = V/R$ and from the graph find the value of V.

(2) The equation connecting two variables R_1 and I is believed to be of the form

$$R_1 = \frac{V}{I} + R_2$$

where V and R_2 are constants. Observed values of R_1 and I are as follows

$$I \ = \ 0.15 \quad 0.23 \quad 0.35 \quad 0.49 \quad 0.71$$
$$R_1 = 72.8 \quad 49.5 \quad 34.7 \quad 26.5 \quad 20.2$$

Using a straight-line graph show that the equation is valid and find values of V and R_2.

(3) The total capacitance C_T of two capacitors connected in parallel is related to the plate separation d of one of them by the equation $C_T = (k/d) + C_s$ where k and C_s are constants. Observed values of C_T and d are as follows

d (mm)	0.1	0.12	0.14	0.16	0.18	0.2
C_s (μF)	15	13.3	12.2	11.25	10.6	10

Using a straight-line graph verify that the equation is correct and hence find the values of k and C_s.

(4) The total resistance R_T of two pieces of wire connected in series is related to the diameter d of one of them by the equation $R_T = (k/d^2) + R_s$ where k and R_s are constants. Using the observed values given below draw a straight-line graph to verify that the equation is correct and from it find the values of k and R_s.

R_T (ohms)	128	34.2	16.9	10.8	8
d (mm)	1	2	3	4	5

(5) The height h of a circular cylinder of constant volume is related to the radius r of its cross-section by the equation $h = k/r^2$ where k is constant. Using the values given plot a linear graph to verify the equation and find the value of k.

h (cm)	0.75	0.19	0.08	0.05	0.03
r (cm)	2	4	6	8	10

(6) Two variable quantities y and x are believed to be related by an equation of the form $(y - b)^2 = ax$. Observed values of y and x are as follows

y	7	9	11	13	15
x	1	4	9	16	25

By plotting a straight-line graph of y against an appropriate function of x verify the equation form and find values for a and b.

(7) The values of two variable quantities p and q were observed to be as follows

p	17.39	27.04	39.69	50.41	72.25
q	13.27	20.03	28.88	36.39	51.68

The variable q is believed to be related to the square root of p by a linear equation. Determine the equation.

(8) The fusing current I of a certain material rolled in the form of a wire of uniform circular cross-section, diameter d, is believed to be dependent on d according to an equation of the form $I = ad^2 + b$ where a and b are constant. Values of I and d are as follows

d (mm)	5	10	15	20	25
I (A)	5.9	9.3	14.9	22.8	32.9

Using a linear graph show that the equation is correct and find the values of a and b.

(9) Observed values of power P and current I in a circuit were as follows

I (A)	0.2	0.4	0.6	0.8	1
P (W)	32	38	48	62	80

Show that the equation connecting P and I is of the form $P = aI^2 + b$ and find the values of the constants a and b.

(10) The volume V of a cylinder of uniform circular cross-section, radius r, is related to the radius by the equation $V = kr^2$. Using the following observed values of V and r for a particular cylinder, draw a straight-line graph and determine the value of the constant k.

V	363.2	694.1	1093.7	1410.4	2270
r	3.4	4.7	5.9	6.7	8.5

(11) The impedance Z of an electrical circuit to alternating current is believed to be related to the frequency f of the current by an equation of the form $Z^2 = k + mf^2$. Observed values of Z and f for a particular circuit are

f (Hz)	10	20	30	40	50
Z (Ω)	111.8	206.2	304.1	403.1	502

By plotting a straight-line graph show that this form of equation is valid and find values of the constants k and m for this circuit.

(12) The equation connecting the inductance L of a coil and the number of turns t is believed to be of the form $L = at^2 + b$ where a and b are constants. Observed values of L and t are as follows

L (H)	28	88	188	328	508
t	5	10	15	20	25

By plotting a linear graph show that the equation form is correct and find the values of a and b.

(13) The anode current I of a diode is related to the anode voltage V by an equation of the form $I = kV^{3/2}$ where k is a constant. Observed values of I and V are

I (mA)	0.64	2.16	5.12	10
V (V)	4	9	16	25

Using a straight-line graph find the value of k.

(14) Two variables x and y are related by the equation $y = kx^n$ where k and n are constants. Using the given values of x and y plot a straight-line graph and determine the values of the constants

x	1	3	4	5	6
y	1.2	2.7	4.8	7.5	10.8

(15) The following values of breakdown voltage V of an insulator of thickness d were observed under test conditions.

d (mm)	3.37	6.86	15.63	21.9
V (kV)	215	345	597.5	749.5

Using a straight-line graph show that V and d are related by an equation of the form $V = kd^n$ where k and n are constants and find the values of k and n.

(16) The values of two variables x and y are as follows

x	1.96	3.24	5.29	6.76	7.84
y	22.14	62.67	196	362.4	522.3

By plotting a straight-line graph show that the equation relating y and x is $y = 3x^{2.5} + 6$.

(17) The pressure and volume of a gas, p and v respectively, are connected by an equation of the form $pv^n = C$ where n and C are constants. Plot a linear graph using the following values and find n and C.

v	4	5	6	7	8
p	73.6	53.6	41.4	33.3	27.5

(18) Show that the variables B and V having values given below are related by an equation of the form $B = aV^b$, where a and b are constants. Find the values of a and b.

V	64	100	144	196	256
B	58.65	179	445.4	962.7	1877

(19) It is known that a variable y is proportional to a variable x raised to a power. Use the following values of x and y to determine the connecting equation.

x	53.18	75.68	138.18	470.9
y	45.97	73.63	164.3	842.64

(20) The total iron loss P in a transformer is related to the frequency f by an equation of the form $P = k_1 f + k_2 f^2$. Values of P and f are

P (W)	5.69	18.25	37.68	64	97 18
f (Hz)	25	50	75	100	125

By plotting a graph of P/f against f find the values of the constants k_1 and k_2.

4 Mensuration

The word 'mensuration' means 'measuring' or 'measurement' and is the word normally applied to the branch of mathematics concerned with the determination of lengths, areas and volumes. These include not only the obviously required details of engineering components but also what appear at first sight to be 'academic' measurements of, for example, the area contained between part of a graph and the axes of the variables. In engineering such areas may be particularly important; for example the area beneath a voltage or current graph, which is proportional to the energy absorbed by the circuit, and the area of the magnetic hysteresis loop (a graph plotting flux density against magnetising force), which again is a measure of the energy required to carry out magnetising and demagnetising. The area beneath the graph of a variable quantity is also used to determine the average value or the equivalent constant value of that quantity — for example, with alternating current, to determine the value of the equivalent direct current (that is, the direct current which would deliver the same power to a circuit).

In the first part of the chapter we shall be concerned with methods of determining the area beneath a curve and the determination of equivalent values of variable quantities. In the second part we shall examine further methods of determining areas and volumes of plane figures and of solids.

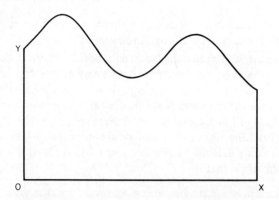

Figure 4.1

THE MID-ORDINATE RULE

The mid-ordinate rule may be used to determine the area of a figure bounded by a curve such as the one shown in figure 4.1 and also to determine the average height of the curve above the base line. If lines OY and OX in the figure represent y and x axes respectively (or axes plotting any other two variables), and the curve is a graph plotting y against x then the mid-ordinate rule may be used to determine the average value (or other equivalent value) of y as x varies between different values.

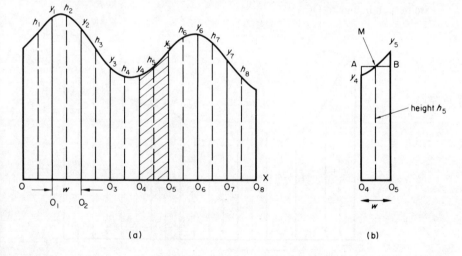

Figure 4.2

Consider the figure shown in figure 4.2a. To apply the mid-ordinate rule to the figure divide the base OX into an equal number of parts each of width w and construct lines $O_1 y_1$, $O_2 y_2$, $O_3 y_3$, etc., perpendicular to OX as shown. These lines are called *ordinates*. Now bisect each strip by the dotted lines of height h_1, h_2, h_3 as shown. These lines are the *mid-ordinates*.

Now consider the strip shown shaded in part a of the figure and drawn separately for clarity in part b. The approximate area of this strip is the product of mid-ordinate × base, that is, $h_5 \times w$. Strictly $h_5 \times w$ is the area of the rectangle $O_4 AB O_5$ but, since the strip area is the area of this rectangle *minus* the area AMy_4 and *plus* the area $My_5 B$ and these two latter areas are approximately the same, the area of the strip and the area of the rectangle are approximately equal.

Returning to figure 4.2a the total area under the curve is equal to the sum of the areas of the eight strips, that is

$$\text{Area} = h_1 w + h_2 w + h_3 w + h_4 w + h_5 w + h_6 w + h_7 w + h_8 w$$

$$= w(h_1 + h_2 + h_3 + h_4 + h_5 + h_6 + h_7 + h_8)$$

or, in words, the area is equal to the strip width × the sum of the mid-ordinates.

It should be noted that this method gives only an approximate result; a more accurate one may be obtained by using Simpson's rule (discussed later) while the most accurate result is obtained by integration (discussed in *Third-year Technician Mathematics and Applications*).

Example 4.1

Find the area of the shape ABCO shown in figure 4.3. AO is parallel to BC and both are perpendicular to OC.

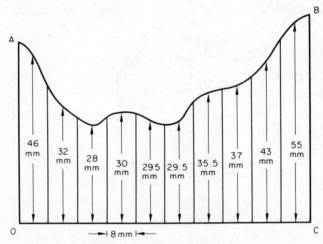

Figure 4.3

Divide the shape as shown into 10 strips of width 8 mm and measure the mid-ordinates. These are 46, 32, 28, 30, 29.5, 29.5, 35.5, 37, 43 and 55 respectively.

The area is 8 (46 + 32 + 28 + 30 + 29.5 + 29.5 + 35.3 + 37 + 43 + 55) mm² that is 8 × 365.5 mm² or 2924 mm².

The student may wonder how many strips to construct in a given figure. Consideration of the theory above indicates that the more strips there are, the more accurate is the result, for then the approximation between the rectangle area and the strip area becomes closer. The answer then is to construct as many strips as conveniently possible.

Average value of a varying quantity

As we have seen in earlier work many engineering quantities are interrelated such that the value of one depends upon the value of another. Common examples include voltages and currents that vary in magnitude with time. Often we are concerned with the average value (or mean value) of the dependent variable — for example, voltage or current — over a range of values of the independent variable — for example, time.

If a graph of a varying quantity is plotted using the dependent variable as ordinate and the independent variable as abscissa, the height of the curve above the horizontal axis represents the value of the varying quantity for any particular value of the variable plotted along the horizontal axis. Thus the average height of the curve above the axis or base line is the average value of the quantity under discussion.

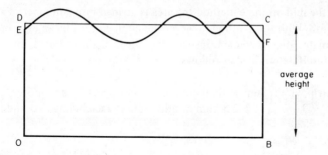

Figure 4.4

Average height of a curve

The average height of a curve above a particular base line is the height of a rectangle having the same area as the area beneath the curve (bounded by the base line and the two outer perpendiculars) and situated on the same base. Thus in figure 4.4 the area bounded by the curve, perpendiculars EO and FB and the base line OB has the same value as the area of rectangle OBCD which is situated on base line OB. The average height is then CB. As shown the height of the curve above OB varies, rising above and falling below the average height CB.

To obtain the average height of any curve (or if the curve is a graph, plotting, say, y against x, the average value of y over a range of values of x) all that is necessary is to divide the area under the curve by the base.

If the mid-ordinate method is used to determine the area under the curve then the area is (approximately) equal to the width of single strip × sum of the mid-ordinates. The base is equal to the width of a single strip × number of strips. Thus

$$\text{average height} = \frac{\text{width of single strip} \times \text{sum of mid-ordinates}}{\text{width of single strip} \times \text{number of strips}}$$

$$= \frac{\text{sum of mid-ordinates}}{\text{number of strips}}$$

and since each strip has one mid-ordinate

$$\text{average height} = \frac{\text{sum of mid-ordinates}}{\text{number of mid-ordinates}}$$

Note that this formula may only be used for equal-width strips, that is, the mid-ordinate method must be applied correctly.

SIMPSON'S RULE FOR AREA DETERMINATION

As with the mid-ordinate method the area is divided into a number of strips of equal width. With this method however the number must be an even number and mid-ordinates are not erected. Instead, the values of the ordinates are measured and the result determined as follows.

$$\text{Area} = \frac{\text{strip width}}{3}\left(\begin{array}{l}\text{Sum of first and last ordinates} + 4 \times \text{sum of even ordinates} \\ + \, 2 \times \text{sum of odd ordinates excluding first and last}\end{array}\right)$$

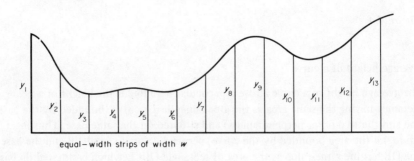

equal—width strips of width *w*

Figure 4.5

For the figure shown in figure 4.5

$$\text{Area} = \frac{w}{3}\, [y_1 + y_{13} + 4\,(y_2 + y_4 + y_6 + y_8 + y_{10} + y_{12}) +$$
$$2\,(y_3 + y_5 + y_7 + y_9 + y_{11})]$$

This result, although approximate, is more accurate than the one obtained by the previous method, though as stated earlier, the most accurate result is obtained by integration; integration does however require a knowledge of the equation of the curve and thus the mid-ordinate and Simpson's rules are most useful where the equation is not known.

Example 4.2

Find the area of the shape ABCO in figure 4.3 using Simpson's rule.

first ordinate length = 50 mm
last ordinate length = 57 mm
even ordinate lengths are 37.5 mm, 29 mm, 27 mm, 36.5 mm, 50 mm
odd ordinate lengths are 29.5 mm, 30 mm, 32.5 mm, 39 mm
strip width is 8 mm

Thus the area is

$$\frac{8}{3}\,[50 + 57 + 4\,(37.5 + 29 + 27 + 36.5 + 50) +$$
$$2\,(29.5 + 30 + 32.5 + 39)] = \frac{8}{3}\,[107 + (4 \times 180) + (2 \times 131)]$$
$$= 2904 \text{ mm}^2$$

ROOT-MEAN-SQUARE VALUES

If we wish to compare one varying quantity with another there are various values of each quantity which can be considered; these include the *instantaneous value*, the *peak or maximum value,* the *average value* and the *root-mean-square value.* The choice of which value to take depends largely on the nature of the quantity and, in particular, the use to which it is being put. In electrical and electronic engineering the varying quantities of most importance are voltage and current, and the most common waveform (that is, a graph of voltage or current against time) is the sine wave shown in figure 4.6a. For such a wave, or any wave that is symmetrical about the time axis, the average value of a complete cycle is zero and no comparison can be made of two voltages or currents having different peak values. Over half a cycle there is an average value and this is used in cases where, for example, half or full-wave rectified voltages and currents are involved (moving-coil instruments, power supplies, etc.).

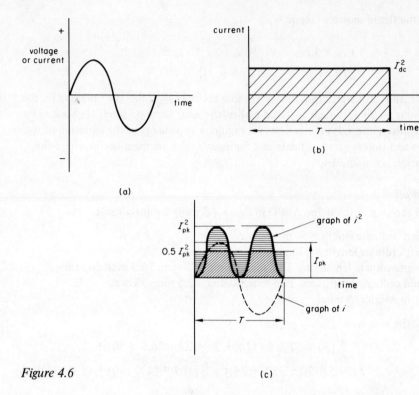

Figure 4.6   (c)

Most frequently we wish to compare voltages and currents from the point of view of their ability to deliver power or energy to a load and in these cases the value of practical importance is the root-mean-square (r.m.s.) value.

Consider a direct current $I_{\text{d.c.}}$ amperes flowing for a time T seconds in a resistance R ohms. The current-time graph is shown in figure 4.6b. Now the power dissipated in the resistance is $I_{\text{d.c.}}^2 R$ and the energy delivered to the resistance is $I_{\text{d.c.}}^2 RT$ joules in T seconds. If a graph of $I_{\text{d.c.}}^2$ against time is plotted as in the figure, the area under the graph over a period of T seconds is the area of the shaded rectangle and equals $I_{\text{d.c.}}^2 T$. Thus the area under the graph is directly proportional to the energy delivered for a constant value of R. This is true for all such graphs whether the current be direct or alternating in nature.

Now consider a sinusoidal alternating current of peak value I_{pk} as shown in figure 4.6c. The graph of the current is shown dotted in the figure and on the same axes a graph of the square of the current is shown by a full line. Note that this graph lies completely above the time axis, that is, it is positive throughout the range T seconds, indicating continuous absorption of power by the resistance. The area under the current-squared graph is proportional to the energy and thus to the power that the current is capable of delivering over the time shown. If now a constant current is taken of value $0.5 \, I_{pk}^2$ the area under this graph is proportional to the energy which a current of value $\sqrt{(0.5 \, I_{pk}^2)}$ can deliver in time T seconds. Examination of such a figure drawn to scale shows that the area under the constant-value graph is equal to the area under the varying-value graph (the cross-hatched areas are common and the horizontally hatched sections above the line $0.5 \, I_{pk}^2$ are of equivalent area to the horizontally hatched sections below the line). Thus the constant value, or direct, current of value $\sqrt{(0.5 \, I_{pk}^2)}$ can deliver the same power as the alternating current of peak value I_{pk}. This value of equivalent d.c. is called the root-mean-square or r.m.s. value since it is the square root of the mean value of the square of the alternating current. The r.m.s. value of currents and voltages is by far the most commonly used value. The equations to remember are

$$I_{r.m.s.} = 0.707 \, I_{pk}$$
$$V_{r.m.s.} = 0.707 \, V_{pk} \, (\text{since } \sqrt{0.5} = 0.707)$$

and

$$I_{pk} = 1.414 \, I_{r.m.s.}$$
$$V_{pk} = 1.414 \, V_{r.m.s.} \quad (\text{since } 1/0.707 = 1.414)$$

provided that the waveforms are sinusoidal.

The methods used so far for obtaining average values can also be employed with equal ease to determine r.m.s. values, the procedure being to find the mean or average of the square of the voltage or current (or other quantity) and then to determine its square root. In general, then

$$\text{r.m.s. value of any quantity} = \sqrt{(\text{mean of the squares of the quantity})}$$

$$= \sqrt{\left(\frac{\text{area under 'squares' graph}}{\text{base}} \right)}$$

Example 4.3

A sinusoidally varying current of peak value 5 A has a periodic time of 20 ms. Using both the mid-ordinate rule and Simpson's rule determine (a) the average value of the current over half a cycle, (b) the r.m.s. value of the current.

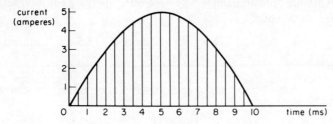

Figure 4.7

One half-cycle of the current waveform is shown in figure 4.7. Although it is not strictly necessary to draw the graph, since values of both ordinates and mid-ordinates can be obtained from tables (as explained below), the figure is included to show the strips into which the area is divided. Ordinates are drawn at time intervals of one millisecond beginning at $t = 0$ (where t is time), through $t = 1$, $t = 2$, etc., to $t = 10$ ms. Mid-ordinates are drawn at time intervals of 1 ms beginning at $t = 0.5$ ms through $t = 1.5$, $t = 2.5$, etc., to $t = 9.5$ ms. As shown, there are ten strips (an even number), so both Simpson's rule and the mid-ordinate rule can be used to find areas.

Values of ordinates and mid-ordinates may be taken from the graph but, since tables must be used in order to draw the graph in the first place, it is equally convenient and more accurate to obtain these values from tables directly. Now, as was explained in *First-year Technician Mathematics*, the equation of any quantity that varies in proportion to the sine of any angle θ is given by

$$y = y_{max} \sin \theta$$

where y_{max} is the maximum value of y. Thus in this case, writing current as i we have

$$i = 5 \sin \theta$$

since 5 A is the maximum value of the current. If a graph of i against θ is plotted, i is normally taken along the vertical (ordinate) axis and θ along the horizontal (abscissa) axis. The angle θ varies in value from $0°$ at the origin to $360°$ at the end of a complete cycle, and so is $180°$ at the end of half a cycle.

If θ were plotted along the horizontal axis in this problem it would have the value

θ (degrees)	0	18	36	54	72	90	108	126	144	162	180
t (ms)	0	1	2	3	4	5	6	7	8	9	10

Thus we can use tables to find i, for when

$$t = 0, \qquad \theta = 0°, \qquad i = 5 \sin 0 = 0$$

and when

$$t = 1, \qquad \theta = 18°, \qquad i = 5 \sin 18 = 0.309$$

These values are the ordinates and will be used when applying Simpson's rule. The mid-ordinates occur when

$$t = 0.5 \qquad 1.5 \qquad 2.5 \qquad 3.5 \qquad \text{etc. ms}$$

that is when

$$\theta = \quad 9 \qquad\quad 27 \qquad\quad 45 \qquad\quad 63 \qquad \text{etc. degrees}$$

and

$$i = \quad 5 \sin 9 \quad 5 \sin 27 \quad 5 \sin 45 \quad 5 \sin 63 \text{ etc. A}$$

Table 4.1 gives the complete set.

Table 4.1

| | Ordinates | | | | | Mid-ordinates | | | |
Time (ms)	Angle (°)	Sine	Current (A)	Current^2	Time (ms)	Angle (°)	Sine	Current (A)	Current^2
0	0	0	0	0	0.5	9	0.1564	0.782	0.6115
1	18	0.3090	1.545	2.387	1.5	27	0.454	2.27	5.153
2	36	0.5878	2.939	8.638	2.5	45	0.7071	3.535	12.5
3	54	0.8090	4.045	16.36	3.5	63	0.891	4.455	19.85
4	72	0.9511	4.755	22.61	4.5	81	0.9877	4.938	24.38
5	90	1	5	25	5.5	99	0.9877	4.938	24.38
6	108	0.9511	4.755	22.61	6.5	117	0.891	4.455	19.85
7	126	0.8090	4.045	16.36	7.5	135	0.7071	3.535	12.5
8	144	0.5878	2.939	8.638	8.5	153	0.454	2.27	5.153
9	162	0.3090	1.545	2.387	9.5	171	0.1564	0.782	0.6115
10	180	0	0	0					

Note that the values of i as θ varies from 0 to $90°$ are repeated in decreasing order as θ varies from $90°$ to $180°$ since the curve is symmetrical about $\theta = 90°$ ($t = 5$ ms). Note also that a column showing current^2 is included in order to obtain the r.m.s. value using the two methods.

(a) Average value over half a cycle. The average value over half a cycle is the area under the curve between $t = 0$ and $t = 10$ ms divided by the base (10 ms). Using the mid-ordinate rule

$$\text{Area under the curve} = \text{width of strip} \times \text{sum of mid-ordinates}$$

The strip width = 1 ms and the sum of the mid-ordinates is obtained by adding the 'current' column in the mid-ordinate section of table 4.1. This equals 31.96. Thus the area equals 31.96×10^{-3} (1 ms = 10^{-3}s) and the average value is

$$\frac{31.96 \times 10^{-3}}{10 \times 10^{-3}}$$

which equals 3.196 A

Alternatively, since the strips are of equal width

$$\text{average value} = \frac{\text{sum of mid-ordinates}}{\text{number of mid-ordinates}}$$

$$= \frac{31.96}{10}$$

$$= 3.196 \text{ A as before}$$

Using Simpson's rule

Area under curve = $\frac{1}{3}$ × strip width $\begin{pmatrix} \text{sum of first and last ordinate} + 4 \times \text{sum of} \\ \text{even ordinates} + 2 \times \text{sum of odd ordinates} \end{pmatrix}$

The first ordinate and the last ordinate are both zero.

$$4 \times \text{sum of even ordinates} = 4\,(1.545 + 4.045 + 5 + 4.045 + 1.545)$$

$$= 4 \times 16.18$$

$$= 64.72$$

$$2 \times \text{sum of odd ordinates} = 2 \times (2.939 + 4.755 + 4.755 + 2.939)$$

$$= 2 \times 15.388$$

$$= 30.776$$

Thus

$$\text{area} = \frac{1}{3} \times 10^{-3} \times (0 + 64.72 + 30.776)$$

$$= \frac{95.496}{3} \times 10^{-3}$$

and since the base is 10×10^{-3}s

$$\text{average value} = \frac{95.496 \times 10^{-3}}{3 \times 10 \times 10^{-3}}$$

$$= 3.183 \text{ A}$$

Note that in both the mid-ordinate and Simpson's method the strip width and base are both taken in the units in which the horizontal axis variable (in this case, time) is measured. This is not actually necessary, because in finding the average the strip width is in the numerator and the base is in the denominator so the units cancel out. Thus the strip width in this problem can be taken as 1 and the base width as 10.

(b) The r.m.s. value may be determined using a half or full cycle of the current^2-time graph. It is not necessary to draw the graph since values of current have already been obtained. The procedure is to find the average of the square of the current and then determine the square root of the average. Using the mid-ordinate rule

$$\text{average value of } i^2 = \frac{\text{sum of } i^2 \text{ mid-ordinates}}{\text{number of mid-ordinates}}$$

$$= \frac{124.989}{10}$$

$$= 12.4989$$

the square root of the average value of i^2 = 3.535, so

r.m.s. value = 3.535

Using Simpson's rule

the sum of first and last ordinates is zero.

sum of even ordinates = 2.387 + 16.36 + 25 + 16.36 + 2.387

= 62.494

sum of odd ordinates = 8.638 + 22.61 + 22.61 + 8.638

= 62.496

therefore average value of $i^2 = \frac{1}{3} \times 1 \times [0 + (4 \times 62.494) + (2 \times 62.496) \div 10$

= 12.499

and

r.m.s. value = $\sqrt{12.499}$

= 5.535 A

It can be shown that the average value of a sine wave over half a cycle is equal to $(2/\pi) \times$ peak value and as shown above the r.m.s. value is equal to 0.707 × peak value. For a peak current of 5 A

average current = $\frac{2}{\pi} \times 5$ = 3.182 A

r.m.s. value = 0.707 × 5 = 3.535 A

A comparison between results follows.

	By formula	Mid-ordinate rule	Simpson's rule
Average	3.182	3.196	3.183
r.m.s.	3.535	3.535	3.533

As can be seen Simpson's rule is the more accurate of the two methods.

The previous example was concerned with a regular curve having a known equation and no graphs need have been drawn. In the next example this is not so.

Example 4.4

The waveform of a transient current in a certain circuit is as shown in figure 4.8. Determine the average and r.m.s. value of this current using the mid-ordinate method.

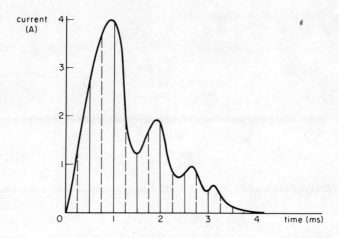

Figure 4.8

The figure is divided into eight strips and the mid-ordinate values are given in table 4.2.

Table 4.2

Strip	Mid-ordinate	Mid-ordinate^2
1	1.2	1.44
2	3.72	13.84
3	1.8	3.24
4	1.7	2.89
5	0.82	0.67
6	0.82	0.67
7	0.35	0.123
8	0.05	0.0025
SUM	10.46	22.88

Average value $= \dfrac{10.46}{8} = 1.307$ A

Average value of squares $= \dfrac{22.88}{8} = 2.86$

Root-mean-square value $= \sqrt{2.86} = 1.69$ A

Summary of values of a varying quantity and their relationships

$$\text{The average value} = \frac{\text{area under graph}}{\text{base}}$$

$$= \frac{2}{\pi} \times \text{peak value for pure sine wave}$$

$$\text{The r.m.s. value} = \frac{\text{area under graph of squares}}{\text{base}}$$

$$= 0.707 \times \text{peak value for pure sine wave}$$

The ratio r.m.s. value: average value is called the *form factor*. For a sine wave this equals

$$\frac{0.707 \times \text{peak}}{(2/\pi) \times \text{peak}} = 1.11$$

If the form factor for any waveform exceeds 1.11 the waveform contains a number of peaks. Examine figure 4.8: from example 4.4 the average value is 1.307 A and the r.m.s. value is 1.69 A; the form factor is thus 1.29. If the form factor is less than 1.11 the waveform tends to have a flat-topped appearance.

The ratio maximum value: r.m.s. value is called the *crest factor* and, from previous work, for a sine wave it is 1.414.

Example 4.5

Determine the average value over half a cycle, the r.m.s. value, the form factor and crest factor of a rectangular voltage waveform alternating about zero and of peak value 100 V.

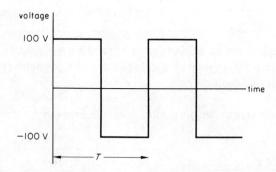

Figure 4.9

The graph of the voltage against time is shown in figure 4.9. Taking T as the periodic time and thus $0.5T$ as the time for half a cycle

Area under graph of voltage = 100 × 0.5T (rectangle)

Base of graph   = 0.5T

Thus

average value $= \dfrac{100 \times 0.5T}{0.5T}$

                              = 100 V

Area under graph of (voltage)^2 = 100^2 × 0.5T

Base of graph   = 0.5T as before

Thus

mean of squares $= \dfrac{100^2 \times 0.5T}{0.5T} = 100^2$

and

r.m.s. value   = 100 V

Thus

form factor $= \dfrac{\text{r.m.s. value}}{\text{average value}} = 1$

and

crest factor $= \dfrac{\text{maximum value}}{\text{r.m.s. value}} = 1$

As can be seen the peak value, average value and r.m.s. values are the same. Note the flat-topped appearance resulting in a form factor of less than that of a pure sine wave (1.11).

Exercise 4.1

(1) Plot the graph of $y = 4x + 3$ between $x = 0$ and $x = 4$. Calculate the area under the graph (between these values of x and the x axis) by geometrical means and also by using the mid-ordinate rule.

(2) Determine the area under the graph $y = x^2 + 3$ between $x = 2$ and $x = 4$ using Simpson's rule.

(3) The depth of a river is measured at two-metre intervals to give the following results

Distance from bank (m)	2	4	6	8	10	12	14	16	18
Depth of river (m)	1.4	1.8	2.1	2.7	2.9	2.8	2.4	2.1	1.2

If the river is moving at 1.1 m/s estimate the volume of water moved per minute.

.(4) The thickness of a resistive film deposited on a ceramic substrate to form a rectangular surface of 2.16 mm X 1.8 mm is determined at 0.27 mm intervals from one edge. The measurements are

Distance from one edge (mm) 0   0.27 0.54 0.81 1.08 1.35 1.62 1.89 2.16
Thickness of film (mm)   0.62 0.63 0.59 0.61 0.58 0.62 0.63 0.62 0.61

Estimate the average thickness of film and the volume of film deposited.

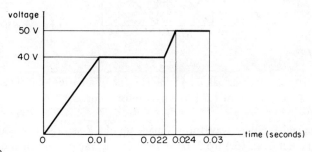

Figure 4.10

(5) The voltage–time graph of a certain voltage is shown in figure 4.10. Determine the average and r.m.s. value of the voltage over the period shown.

(6) Draw the graph of $i = 10 \sin 314t$ between $t = 0$ and $t = 20$ ms and determine the r.m.s. value of the current i (a) by the mid-ordinate rule   (b) by the use of the formula.

(7) Draw the graph of $y = 2x^2 + 1$ between $x = 0$ and $x = 3$ and determine the average value of y over this range of x using Simpson's rule.

(8) The current flowing in a 100 Ω resistor varies as the graph shown in figure 4.11. Estimate the energy absorbed by the resistor during the period shown.

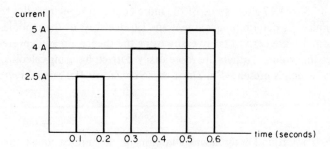

Figure 4.11

(9) Determine the average value, r.m.s. value, form factor and crest factor of the voltage having a waveform shown in figure 4.12.

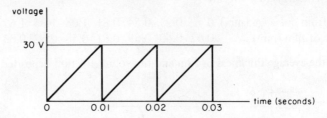

Figure 4.12

(10) Calculate the average value and r.m.s. value of a current having a form factor of 1.2 and a crest factor of 1.7 if the maximum value is 3.7 A.

(11) A triangular voltage waveform is made up of a linear rise from 0 to 15 V in 20 ms and a fall from 15 V to 0 in the next 10 ms, the pattern then being repeated. Calculate the r.m.s. value of this voltage.

(12) Compare the results obtained for the r.m.s. value of a sinusoidal current having the equation $i = 13 \sin 628t$ when these results are obtained (a) by Simpson's rule (b) by mid-ordinate method   (c) by calculation.

(13) Find the mean and r.m.s. value of a voltage that rises linearly from 0 to 24 V in 0.1 s then falls to zero in the next 0.01 s, the cycle then repeating itself.

(14) Use Simpson's rule to find the average value of a sinusoidal voltage of peak value 15 V and periodic time 40 ms over the first 15 ms of each cycle.

AREAS AND VOLUMES OF REGULAR BODIES

Figures 4.13 to 4.17 show some of the more common regular bodies and summarise the formulae for determining volumes and surface areas of these bodies. Some of these formulae are derived fairly easily using the theory already covered in this and the preceding volume; others are more easily derived by using calculus, the basic theory of which is presented in *Third-year Technician Mathematics and Applications*.

Spheres

A sphere is any round body having all points on the surface equally distant from the centre. The cross-section of a sphere, wherever it is taken, is always a circle and, if taken through the centre, the radius of the circle is the radius of the sphere, that is, the distance between the centre and any point on the surface. Denoting the sphere radius by r, the volume of a sphere may be shown to equal $\frac{4}{3}\pi r^3$ and the surface area $4\pi r^2$.

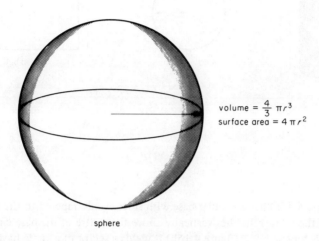

volume = $\frac{4}{3}\pi r^3$
surface area = $4\pi r^2$

sphere

Figure 4.13

Cylinders

If a rectangle is rotated about one of its sides the shape produced by the other parallel side is a cylinder. The length of either of these sides is the cylinder height, shown as h in figure 4.14, and the length of either of the remaining sides of the rectangle is the radius of the circle that forms each end of the cylinder. A cross-section of a cylinder taken in a plane parallel to the height is a rectangle and a cross-section taken in a plane perpendicular to the height is a circle. Since

$$\text{volume} = \text{height} \times \text{area of circular cross-section}$$

$$\text{volume} = \pi r^2 h$$

where r is the radius of the circle. The total surface area is 2 × area of each circular end-section added to the curved surface area. If the cylinder is 'opened out' the curved surface area forms a rectangle of sides $2\pi r$ and h. The total area is thus $2\pi r^2 + 2\pi rh$. (See figure 4.14 and also *First-year Technician Mathematics* for further explanation.)

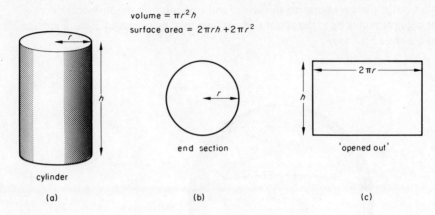

volume = $\pi r^2 h$
surface area = $2\pi r h + 2\pi r^2$

end section

'opened out'

cylinder

(a)   (b)   (c)

Figure 4.14

Cones

A cone (figure 4.15a) has a circular base with sides evenly tapered to the vertex (the 'top' in the figure) that lies vertically above the centre of the base circle. If the curved area is 'opened out' (figure 4.15b) it yields a sector of a circle having a radius equal to the slant side of the cone (*l* in the figure) and an arc equal to the circumference of the base circle ($2\pi r$ in the figure). The sector of the circle, radius *l*, that makes the curved surface has an area in direct proportion to the angle θ in the figure.

$$\text{Sector area} = \frac{\theta}{2\pi} \times \pi l^2 \quad \text{(where } \theta \text{ is in radians)}$$

volume = $\frac{1}{3}\pi r^2 h$
curved surface area = $\pi r l$
total area = $\pi r l + \pi r^2$

cone   'opened out'

(a)   (b)

Figure 4.15

that is, a fraction $\theta/2\pi$ of the total circle area πl^2 (if $\theta = 2\pi$ the sector would become a complete circle).

$$\text{Arc length } 2\pi r = l\theta \quad \text{(if } \theta \text{ is in radians*)}$$

Thus

$$\theta = \frac{2\pi r}{l}$$

and

$$\text{cone curved surface area} = \text{sector area} = \frac{(2\pi r)}{l} \times \frac{1}{2\pi} \times \pi l^2$$

$$= \pi r l$$

$$\text{base circle area} = \pi r^2$$

therefore

$$\text{total surface area of cone} = \pi r^2 + \pi r l$$

It can be shown that the cone volume $= \frac{1}{3} \pi r^2 h$ as indicated.

The frustrum of a cone is shown in figure 4.16a. It consists of a cone section formed by slicing the cone parallel to the base. Now

$$\text{curved surface area of total cone} = \pi R L$$

and

$$\text{curved surface area of cone above section} = \pi r (L - l)$$

thus

$$\text{curved surface area of frustrum} = \pi R L - \pi r L + \pi r l$$

$$= \pi L (R - r) + \pi r l$$

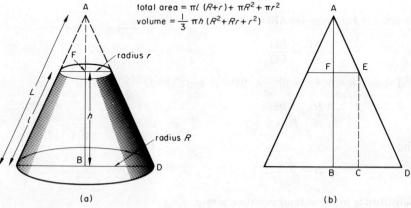

total area $= \pi l (R+r) + \pi R^2 + \pi r^2$

volume $= \frac{1}{3} \pi h (R^2 + Rr + r^2)$

(a)

(b)

Figure 4.16

*See First-year Technician Mathematics, p. 150.

If we examine the figure 4.16b which is a cross-section of the total cone taken perpendicular to the base, AD = L, ED = l, FE = BC = r and BD = R. $\triangle$ABD and $\triangle$ECD are similar triangles and so

$$\frac{ED}{AD} = \frac{CD}{BD}$$

that is

$$\frac{l}{L} = \frac{R-r}{R}$$

so

$$R - r = \frac{Rl}{L}$$

and thus frustrum curved surface area = $\pi L \left(\frac{Rl}{L}\right) + \pi rl$

$$= \pi Rl + \pi rl$$

$$= \pi l (R + r)$$

and total surface area, which equals curved surface + end sections,

$$= \pi l (R + r) + \pi R^2 + \pi r^2$$

The volume of the frustrum is equal to the difference between the volume of the total cone and the volume of the cone above the section. That is

$$\text{volume of frustrum} = \tfrac{1}{3}\pi R^2 H - \tfrac{1}{3}\pi r^2 (H - h)$$

where H is the height of the total cone (AB in figure 4.16b)

$$= \tfrac{1}{3}\pi H (R^2 - r^2) + \tfrac{1}{3}\pi r^2 h$$

In figure 4.16b triangles ABD and ECD are similar so that

$$\frac{AB}{EC} = \frac{BD}{CD}$$

and since AB = H, EC = FB = h, BD = R and CD = $R - r$, we have

$$\frac{H}{h} = \frac{R}{R-r}$$

and

$$H = \frac{Rh}{R-r}$$

Substituting in the volume equation above

$$\text{volume} = \tfrac{1}{3}\pi \, \frac{Rh}{R-r} (R^2 - r^2) + \tfrac{1}{3}\pi r^2 h$$

$R^2 - r^2$ is the difference between two squares and is equal to $(R - r)(R + r)$, therefore

$$\text{volume} = \frac{1}{3}\pi Rh\,(R + r) + \frac{1}{3}\pi r^2 h$$

$$= \frac{1}{3}\pi h\,(R^2 + rR + r^2)$$

The technique of using similar triangles as illustrated is often most useful with problems involving frustrums of cones and it will be used again later.

Pyramids

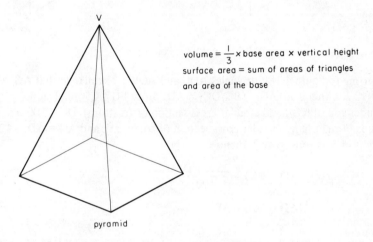

$$\text{volume} = \frac{1}{3} \times \text{base area} \times \text{vertical height}$$

surface area = sum of areas of triangles and area of the base

pyramid

Figure 4.17

A pyramid is a figure having a polygon as its base, the sides of the polygon forming the bases of triangular sides that meet at a common vertex. Figure 4.17 shows a pyramid with a square base and four triangular shaped sides meeting at vertex V.

For all pyramids it can be shown that the total volume is $\frac{1}{3} \times$ base area $\times$ vertical height. The surface area may normally be determined by finding the area of each side and the area of the base and adding.

The following examples illustrate the use of the above formulae.

Example 4.6

Figure 4.18a shows a cathode ray tube. It consists of a cylinder of length 80 mm and diameter 50 mm connected to the frustrum of a cone of base diameter 200 mm. The height of the frustrum is 120 mm. Find the height of the complete cone and calculate (a) the total surface area of the tube (b) the total volume of the tube.

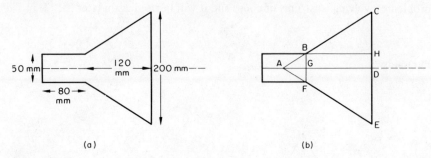

Figure 4.18

Referring to figure 4.18b, triangles ADC and AGB are similar so that AG/AD = BG/DC and therefore AG = (BG/DC) × AD. Since BG = 25 mm (radius of cylinder and of left-hand circular cross-section of frustrum), DC = 100 mm (radius of right-hand circular cross-section of frustrum), and AG = AD − 120 mm (height of frustrum, given), then

$$AD - 120 = \frac{25 \times AD}{100}$$

$$4AD - 480 = AD$$

therefore

$$\text{Cone height AD*} = \frac{480}{3} = 160 \text{ mm}$$

To find the slant side: The slant side of the frustrum is BC and its length may be determined using △CBH in which

$$BH = GD = 120 \text{ mm (given) and } CH = CD - HD = 100 - 25 = 75 \text{ mm}$$

Then

$$BC^2 = BH^2 + CH^2$$

$$= 120^2 + 75^2$$

and

$$BC = 141.5 \text{ mm}$$

*Note the word 'height' means the perpendicular distance between the base and vertex of the cone. Strictly speaking, since this cone is lying on its side the word height is incorrect.

To find the area

Area of cylindrical portion = area of circular end + area of curved surface

$$= (\pi \times 25^2) + (2\pi \times 25 \times 80)$$

$$= 1964 + 12\,571$$

$$= 14\,535 \text{ mm}^2$$

Area of frustrum $= \pi \times 141.5 \times (100 + 25) + (\pi \times 100^2)$

$$= 55\,589 + 31\,420$$

$$= 87\,009$$

and

Total area $= 14\,535 + 87\,009 = 101\,544 \text{ mm}^2$

(Note that the left-hand cross-sectional area of the frustrum is not included.)

To find the volume

Volume of cylinder $= \pi \times 25^2 \times 80 = 157.14 \times 10^3 \text{ mm}^3$

Volume of frustrum $= \frac{1}{3}\pi \times 120 \,(100^2 + 100 \times 25 + 25^2)$

$$= 1650 \times 10^3 \text{ mm}^3$$

therefore total volume

Total volume $\quad = (157.14 + 1650) \times 10^3$

$$= 1807.14 \times 10^3 \text{ mm}^3$$

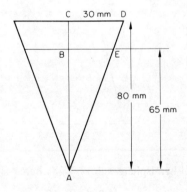

Figure 4.19

Example 4.7

An upturned hollow cone of height 80 mm and diameter 60 mm is filled with liquid to a depth of 65 mm. Find the diameter of the liquid surface.

A diagram of the cone is shown in figure 4.19. Since triangles ACD and ABE are similar BE/CD = AB/AC and since CD = 30 mm, AB = 65 mm and AC = 80 mm

$$\frac{BE}{30} = \frac{65}{80}$$

therefore

$$BE = \frac{30 \times 65}{80}$$

$$= 24.4 \text{ mm}$$

and

$$\text{diameter of surface} = 2 \times 24.4$$

$$= 48.8 \text{ mm}$$

Example 4.8

A solid that is made up of a cone and a hemisphere has the dimensions shown in figure 4.20. Find the total surface area of the solid.

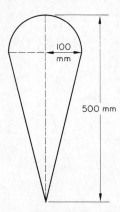

100 mm

500 mm

Figure 4.20

$$\text{Surface area of hemisphere} = 2\pi r^2 \qquad \text{(where } r \text{ is the radius)}$$

$$= 2 \times \pi \times 10^4$$

$$= 628.4 \times 10^2 \text{ mm}^2$$

To find the surface area of the cone the slant-side length is required.

$$\text{Height of cone} = 500 - 100 \text{ mm}$$

$$= 400 \text{ mm}$$

Radius óf base of cone = 100 mm, therefore

$$\text{slant-side length} = \sqrt{(400^2 + 100^2)}$$

$$= 412.3 \text{ mm}$$

$$\text{Surface area of cone} = \pi \times 100 \times 412.3$$

$$= 1295.8 \times 10^2 \text{ mm}^2$$

therefore

$$\text{Total area} = (628.4 + 1295.8) \times 10^2$$

$$= 1924.2 \times 10^2 \text{ mm}^2$$

Example 4.9

The volume of a sphere radius r is to be twice that of a cone having the same base radius. Find an equation relating the cone height h to the base radius.

$$\text{Sphere volume} = \frac{4}{3}\pi r^3$$

$$\text{Cone volume} = \frac{1}{3}\pi r^2 h$$

therefore

$$\frac{4}{3}\pi r^3 = 2 \times \frac{1}{3}\pi r^2 h$$

then

$$2r = h$$

and

$$r = \frac{h}{2}$$

Example 4.10

A cylindrical rod of length 1 metre and diameter 25 mm is to be tapered to give a cross-sectional diameter of 20 mm at one end and 18 mm at the other. Find the volume of metal removed during machining.

The volume of metal removed is the difference between the volume of a cylinder, length 1 m, diameter 25 mm, and the volume of the frustrum of a cone of radii 10 mm and 9 mm and height 1 metre.

Volume of cylinder = $\pi r^2 h$ (where r is radius and h is height)

$$= \pi \times 12.5^2 \times 10^{-6} \times 1 \quad \text{m}^3$$

$$= 490.94 \times 10^{-6} \text{ m}^3$$

Volume of frustrum = $\frac{1}{3} \times \pi \times 1 \times$

$$[\ (10 \times 10^{-3})^2 + (10 \times 10^{-3} \times 9 \times 10^{-3}) + (9 \times 10^{-3})^2\ ]$$

$$= \frac{\pi}{3} \times 2.71 \times 10^{-4} \text{ m}^3$$

$$= 2.8383 \times 10^{-4} \text{ m}^3$$

Volume removed $= (4.9094 \times 10^{-4}) - (2.8383 \times 10^{-4})$

$$= 2.0711 \times 10^{-4} \text{ m}^3$$

Example 4.11

Calculate the volume and total surface area of a pyramid having 4 sides, each one consisting of an equilateral triangle of side 100 mm.

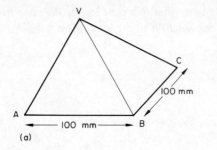

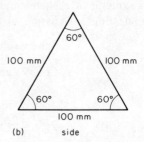

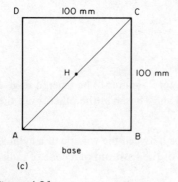

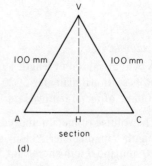

Figure 4.21

Figure 4.21a shows the pyramid, each side (figure 4.21b) and its base (figure 4.21c). The height of each triangle is given by $\sqrt{(100^2 - 50^2)}$, since the height is the length of a line perpendicular to the base and bisecting it into two 50 mm lengths.

$$\text{Height} = \sqrt{7500}$$

$$= 86.6 \text{ mm}$$

$$\text{Area} = \frac{1}{2} \times \text{base} \times \text{height}$$

$$= \frac{1}{2} \times 100 \times 86.6$$

$$= 4330 \text{ mm}^2$$

$$\text{Total area} = 4 \times 4330 + 100^2 \quad (4 \times \text{side area} + \text{base area})$$

$$= 17320 + 10000$$

$$= 273.2 \times 10^2 \text{ mm}^2$$

$$\text{Volume} = \frac{1}{3} \times \text{base area} \times \text{vertical height}$$

The vertical height is the length of the dotted line from V perpendicular to AC in the section shown in figure 4.21d, that is, VH.

$$\text{Length AC} = \sqrt{(AB^2 + BC^2)}$$

$$= \sqrt{(100^2 + 100^2)}$$

$$= 141.4 \text{ mm}$$

therefore

$$HC = AC/2 = 70.7 \text{ mm}$$

and

$$VH = \sqrt{(VC^2 - HC^2)}$$

$$= \sqrt{(100^2 - 70.7^2)}$$

$$= 70.7 \text{ mm}$$

therefore

$$\text{Volume} = \frac{1}{3} \times 100^2 \times 70.7 = 235.67 \times 10^3 \text{ mm}^3$$

Exercise 4.2

(1) A water tank consisting of a cylinder 0.9 m in diameter and 1.2 m high surmounted by a hemisphere of the same diameter as the cylinder is emptied in 45 seconds. Determine the rate of flow of water in cubic centimetres per minute.

(2) A sand hopper has the form of an inverted cone of height 3.2 m and base diameter 2.1 m. Determine the volume of sand in the hopper when it is two-thirds full.

(3) Determine the rise in the liquid level when a sphere of diameter 100 cm is immersed in a cylindrical water-tank of cross-sectional area 0.45 m^2.

(4) Find the volume of a solid tapered plug of length 12 cm and diameters 1.5 cm and 2.5 cm. Hence determine the waste material if such a plug is turned from a cylindrical rod of diameter 2.5 cm and length 12 cm.

(5) An integrated circuit contained in a cylindrical can of height 8 mm and diameter 5 mm is mounted on a rectangular heat sink of dimensions 45 mm X 65 mm. Calculate the total cooling area including the can sides.

(6) Determine the volume of a right pyramid 7 cm high and having a pentagonal base of side 3 cm.

(7) Ten cylindrical ingots of diameter 12 cm and length 15 cm are melted down and recast in the form of a solid cone 25 cm high. Determine the length of the slant side of the cone.

(8) Determine the area of the largest sector of a circle that can be cut from a square of sheet metal of side 20 cm. If this sector is now shaped into a cone calculate the cone height and base diameter.

(9) Determine the side length of a solid cube that has the same volume as a sphere of radius 4 cm.

(10) The sides of a pyramid consist of four equilateral triangles of side 5 cm. Determine (a) the total surface area including the base   (b) the pyramid volume.

(11) A tapered plug of diameters 12 mm and 25 mm and length 4 cm is inserted in a hole of diameter 15 mm. Determine the depth of penetration of the plug into the hole and the fraction of the total volume remaining outside the hole.

(12) When 12 similar ball-bearings are placed inside a cylindrical container of height 50 cm which is two-thirds full of oil, the level of liquid rises by 5 per cent of the original level. Find the diameter of each ball-bearing.

(13) Liquid from a spherical container 4.3 m in diameter is discharged at the rate of 0.25 cubic metres per minute into a second spherical container of diameter 1.5 m. Calculate the time taken for the second container to fill and the percentage volume remaining in the first container.

(14) A circular piece of material of diameter 30 cm is cut into two pieces—the first, a sector of the circle, being used to form a cone of base radius 5 cm. Determine the area of the second piece of the circle.

5 Trigonometry

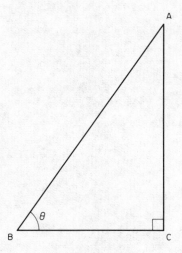

Figure 5.1

In *First-year Technician Mathematics* the following functions of an angle θ lying between 0 and 90°, as in figure 5.1, were given: sine θ = AC/AB; cosine θ = BC/AB; tangent θ = AC/BC (abbreviated sin θ, cos θ and tan θ respectively). Three additional functions are

$$\text{cosecant } \theta \ = \frac{AB}{AC} \quad \text{(reciprocal of sin } \theta\text{)}$$

$$\text{secant } \theta \ = \frac{AB}{BC} \quad \text{(reciprocal of cos } \theta\text{)}$$

$$\text{cotangent } \theta = \frac{BC}{AC} \quad \text{(reciprocal of tan } \theta\text{)}$$

These are abbreviated cosec θ, sec θ and cot θ respectively. They are used in preference to writing $1/\sin \theta$, $1/\cos \theta$ or $1/\tan \theta$ whenever this is more convenient. The values of these functions may be obtained directly from tables of sines, cosines, tangents and reciprocals.

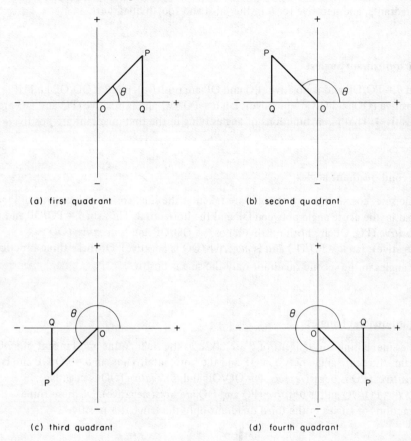

(a) first quadrant

(b) second quadrant

(c) third quadrant

(d) fourth quadrant

Figure 5.2

FUNCTIONS OF ANGLES GREATER THAN 90°

Figure 5.2 shows angles lying (a) in the first quadrant, (b) in the second quadrant, (c) in the third quadrant and (d) in the fourth quadrant. The first quadrant contains angles lying between 0 and 90°, the second contains angles between 90° and 180°, the third contains angles between 180° and 270° and the fourth contains angles between 270° and 360°. As with graphs, lengths above the horizontal line are considered positive, lengths below are considered negative.

Lengths to the right of the vertical line are considered positive and those to the left are considered negative. The length OP is always taken as positive. Thus OQ is positive when θ is in the first and fourth quadrants, and is negative for θ in the second and third quadrants; and PQ is positive in the first and second quadrants, and negative for θ in the third and fourth quadrants.

First-quadrant angles

$\sin \theta$ = PQ/OP and is positive (PQ and OP are positive); $\cos \theta$ = OQ/OP and is positive (OQ and OP are positive); $\tan \theta$ = PQ/OQ and is positive (PQ and OQ are positive). That is, *all* functions of angles lying in the first quadrant are positive.

Second-quadrant angles

The sine, cosine and tangent of θ are taken as the sine, cosine and tangent of $P\hat{O}Q$. that is, the acute angle between OP and the horizontal. Thus $\sin \theta$ = PQ/OP and is *positive* (PQ, OP are both positive); $\cos \theta$ = OQ/OP and is *negative* (OQ is negative); $\tan \theta$ = PQ/OQ and is *negative* (OQ is negative). Of these three functions of angles in the second quadrant only the sine is positive.

Third-quadrant angles

The sine, cosine and tangent of θ are taken as the sine, cosine and tangent of $P\hat{O}Q$; again, the acute angle between OP and the horizontal. Thus $\sin \theta$ = PQ/OP and is *negative* (PQ is negative); $\cos \theta$ = OQ/OP and is *negative* (OQ is negative); $\tan \theta$ = PQ/OQ and is *positive* (PQ and OQ are both negative). Of these three functions of angles in the third quadrant only the tangent is positive.

Fourth-quadrant angles

The sine, cosine and tangent of θ are taken as the sine, cosine and tangent of $P\hat{O}Q$, once again the acute angle between OP and the horizontal. Thus $\sin \theta$ = PQ/OP and is *negative* (PQ is negative); $\cos \theta$ = OQ/OP and is *positive* (OQ, OP are positive); $\tan \theta$ = PQ/OQ and is *negative* (PQ is negative). Of these three functions of angles in the fourth quadrant only the cosine is positive.

A summary of positive functions for quadrants one to four is *all-sin-tan-cos*. (A common mnemonic to aid the memory is '*all stations to Crewe*' which gives the initial letters of the positive functions, that is *A-S-T-C*).

Sometimes an angle may exceed 360°, a common example being when graphs of trigonometrical functions are plotted. However large the angle θ becomes the line OP in the diagram will always lie in one of the quadrants and the above rules always apply. The following examples illustrate how the functions are determined.

Example 5.1

Find the values of the following

(1) sin 145° 12'	(2) cos 289° 13'	(3) tan 221° 3'
(4) sin 357°	(5) tan 108° 51'	(6) cos 384° 34'
(7) tan 547° 32'	(8) cos 722°	(9) tan 914°
(10) sec 310°	(11) cosec 281°	(12) cot 112°

(1) To find sin 145° 12'. This angle lies in the second quadrant. The acute angle made with the horizontal is 180° − 145° 12' = 34° 48' and in the second quadrant the sine is positive, therefore sin 145° 12' = sin 34° 48' = 0.5707 from tables.

(2) To find cos 289° 13'. This angle lies in the fourth quadrant so the cosine is positive and is the cosine of 360° − 289° 18' since this is the acute angle made with the horizontal. Cos 289° 13' = cos 360° − 289° 13' = cos 70° 47' = 0.3291 from tables.

(3) To find tan 221° 3'. This angle lies in the third quadrant and the tangent is positive and equals the tangent of 221° 3' − 180° since this is the acute angle made with the horizontal. Tan 221° 3' = tan 221° 3' − 180° = tan 41° 3' = 0.8709 from tables.

(4) To find sin 357°. This angle lies in the fourth quadrant and the sine is negative. The acute angle is 360° − 357° that is, 3°. Sin 357° = − sin 3° = − 0.0523 from tables.

(5) To find tan 108° 51'. This angle lies in the second quadrant and the tangent is negative. The acute angle is 180° − 108° 51', that is, 71° 9'. Tan 108° 51' = − tan 71° 9' = − 2.93 from tables.

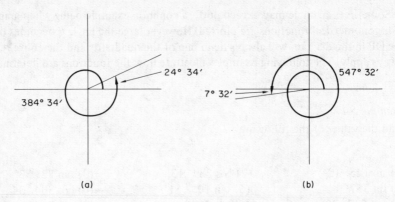

Figure 5.3

(6) To find cos 384° 34'. To make an angle of 384° 34' the line creating the angle (figure 5.3a) has completed one revolution, that is, 360° plus 24° 34'. This lies in the first quadrant where the cosine is positive. Cos 384° 34' = cos 24° 34' = 0.9095 from tables.

(7) To find tan 547° 32'. This angle is equal to 360° + 187° 32', so it lies in the third quadrant (see figure 5.3b.) The acute angle is 187° 32' − 180°, which is 7° 32' and the tangent is negative. Tan 547° 32' = tan (547° 32' − 360°) = tan 187° 32' = − tan 7° 32' = − 0.1323 from tables.

(8) To find cos 722°. The arm has rotated twice to give 360° + 360° and stops at an angle 2° in the first quadrant (see figure 5.4). Cos 722° = cos 2° = 0.9994 from tables.

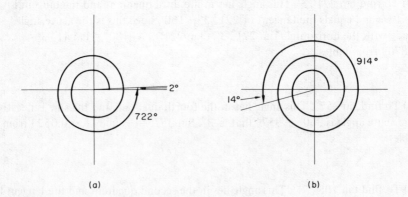

Figure 5.4

(9) To find tan 914°. As shown in figure 5.4b the arm has rotated twice to give 360° + 360° then a further 194° to stop in the third quadrant giving an acute angle of 14°. The tangent is positive in this quadrant. Tan 914° = tan (914° − 720°) = tan 194° = tan (194° − 180°) = tan 14° = 0.2493 from tables.

(10) To find sec 310°. Sec 310° = 1/cos 310° = 1/cos 50° = 1/0.6428 = 1.5557 from tables.

(11) To find cosec 281°. Cosec 281° = 1/sin 281° = 1/[− sin (360° − 281°)] = 1/− sin 79° = 1/− 0.9816 = − 1.0187 from tables.

(12) To find cot 112°. Cot 112° = 1/tan 112° = 1/− tan 68° = 1/− 2.475 = − 0.4040 from tables.

Exercise 5.1

Determine the following

(1) sin 117° 30′	(2) cos 241° 4′	(3) tan 302° 5′	(4) cosec 117° 30′
(5) sec 241° 4′	(6) cot 302° 5′	(7) sin 192° 4′	(8) cos 235° 56′
(9) tan 101°	(10) cos 350°	(11) sin 271°	(12) cot 143°
(13) sin 384°	(14) cos 592°	(15) tan 972° 14′	(16) sec 902°
(17) cosec 413°	(18) cot 1322°	(19) sin 811°	(20) cos 2783°

Trigonometric functions of angles between quadrants

So far we have considered functions of angles falling in one or other of the four quadrants. Consider again figure 5.2. In the first quadrant as θ becomes progressively smaller, approaching zero, length OP becomes closer in value to length OQ and PQ becomes smaller and smaller. Now sin θ = PQ/OP and will reduce in value as PQ becomes smaller, similarly tan θ, which equals PQ/OQ, will reduce in value and cos θ, which equals OQ/OP, will approach unity as OP approaches OQ. Thus at 0° sin θ is zero, cos θ is 1 and tan θ is zero.

As θ approaches 90° (at the junction of the first and second quadrants) OQ becomes smaller and approaches zero, OP and PQ become closer to each other in value. Consequently sin θ approaches unity, cos θ approaches zero and tan θ becomes larger and larger (since the denominator OQ of tan θ, is getting smaller and smaller). At 90°, therefore, sin θ is unity, cos θ is zero and tan θ is infinitely high (written as ∞).

Consider now the second quadrant. As θ approaches 180° OP and OQ again become closer in value and PQ becomes smaller and smaller. On this occasion OQ is negative (OP is always taken as positive). Thus cos θ approaches − 1, and sin θ and tan θ approach zero. At θ = 180°, cos θ = − 1, sin θ = tan θ = 0.

Considering the third quadrant, as θ approaches $270°$ OP and PQ become closer in value (on this occasion PQ being negative) and OQ becomes smaller. Thus sin θ approaches -1, cos θ approaches zero and tan θ increases in value (its value being negative). At $\theta = 270°$, sin $\theta = -1$, cos $\theta = 0$ and tan θ is an infinitely large negative number (written $-\infty$).

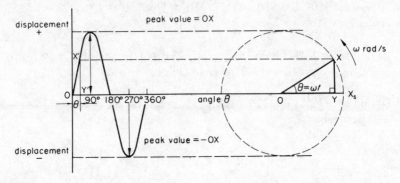

Figure 5.5

EQUATIONS OF SINUSOIDAL FUNCTIONS

In *First-year Technician Mathematics* graphs were plotted of sin θ against θ by taking the displacement of the end of a rotating arm from a horizontal reference axis as being proportional to sin θ, the angle θ being the angle the arm made with the horizontal. This is shown again in figure 5.5. In this figure OX is the rotating arm (fixed at O) and XY is the displacement of X from the horizontal axis. As OX rotates XY changes in magnitude according to the relationship

$$XY = OX \sin \theta \quad \text{(since sin } \theta = XY/OX)$$

and plotting the value of XY against θ results in the sine wave shown. For the value of θ given, XY projects across to give X'Y' on the graph. When $\theta = 90°$ the projection on the graph is the maximum value and equals OX since here OX = XY. Similarly when $\theta = 270°$, the projection is the negative maximum and equals $-$ OX. At $\theta = 0°$ and $180°$ the projection is zero. As indicated, sin θ is positive for values of θ between $0°$ and $180°$ and negative for values of θ between $180°$ and $360°$. This is in agreement with the earlier work in this chapter.

Suppose now that the arm OX is rotated at an angular velocity of ω radians per second (ω is pronounced omega). At any time t seconds after OX begins rotating the arm will have moved through an angle ωt radians. Thus if θ is the angle as shown on the diagram and $t = 0$ when OX lies along the horizontal in the position OX_s then $\theta = \omega t$ and therefore we can write

$$XY = OX \sin \omega t$$

Suppose one revolution of OX takes T seconds so that one complete cycle of the projected wave (from zero to maximum to zero to negative maximum to zero) is drawn in T seconds. T is called the *periodic time* or *period* of the sine wave and, since one revolution is equal to 2π radians ($360°$), then the angle ωT turned through in one revolution is given by

$$\omega T = 2\pi$$

so

$$T = \frac{2\pi}{\omega}$$

The number of cycles that occur in one second of any regularly occurring waveform such as this is called *the frequency*, denoted by f. The unit of f is the cycle per second given the special name the *hertz*, abbreviated Hz. Clearly, if f cycles occur in 1 second then fT cycles occur in T seconds but T seconds is the time taken for one cycle so $fT = 1$, that is

$$T = \frac{1}{f}$$

But

$$T = \frac{2\pi}{\omega} \quad \text{(from above)}$$

therefore

$$\frac{1}{f} = \frac{2\pi}{\omega}$$

and

$$\omega = 2\pi f$$

Thus we can write, for the case shown, $XY = OX \sin 2\pi ft$ where f is the frequency (Hz) and t is time (seconds).

Now XY is the value of the displacement of X from the horizontal at any instant t seconds after the start, that is, XY is the instantaneous value of the variable quantity being plotted. As we have seen, OX is the maximum value so we can write

instantaneous value = maximum value $\times \sin (2\pi \times$ frequency $\times$ time$)$

This relationship is true for all sinusoidally varying quantities and is encountered on innumerable occasions in electrical engineering; typical quantities include voltage, current, electric and magnetic flux density and field strength.

Example 5.2

The instantaneous value i of an alternating current is given by

$$i = 10 \sin 314.2t \text{ amperes}$$

Determine (a) the peak value,   (b) the r.m.s. value,   (c) the period,   (d) the frequency,   (e) the current after 15 milliseconds.

(a) The peak value is 10 A (from the general equation).

(b) The r.m.s. value is $0.707 \times 10 = 7.07$ A (see chapter 4).

(c) From the previous work $2\pi f = 314.2$ where f is frequency (Hz), therefore

$$\text{periodic time} = \frac{1}{f} = \frac{2\pi}{314.2} = 20 \text{ ms}$$

(d) The frequency $f = 10^3/20$ (from part c) = 50 Hz.

(e) Current after 15 milliseconds

$$= 10 \times \sin (314.2 \times 15 \times 10^{-3}) \text{ radians}$$
$$= 10 \times \sin (4.713 \text{ rad})$$
$$= 10 \times \sin 270° = -10 \text{ A (since } \sin 270° = -1)$$

Example 5.3

The r.m.s. value of a sinusoidally changing voltage is 23 V. The periodic time is 4.7×10^{-3} seconds. Obtain an equation for the instantaneous value of this voltage and find its value after 2×10^{-3} seconds.

Since the r.m.s. value is 23 V and the voltage is sinusoidal, the peak value is 23×1.414, that is, 32.52 V. Then periodic time = 4.7×10^{-3} and so frequency = $10^3/4.7 = 212.77$ Hz. The equation is therefore

$$\text{instantaneous value} = 32.52 \sin (2\pi \times 212.77t)$$

where t is in seconds. After 2×10^{-3} seconds

$$\text{voltage} = 32.52 \sin (2\pi \times 212.77 \times 2 \times 10^{-3})$$

The angle is in radians and since 1 rad = $360/(2\pi)$ degrees, then

$$2\pi \times 212.77 \times 2 \times 10^{-3} \text{ rad}$$
$$= \frac{2\pi \times 212.77 \times 2 \times 10^{-3} \times 360}{2\pi} \text{ degrees}$$
$$= 153.19°$$
$$= 153° \, 11'$$

now

$$\sin 153° \; 11' = \sin 26° \; 49'$$

$$= 0.4512$$

therefore

$$\text{voltage} = 32.52 \times 0.4512$$

$$= 14.67 \text{ V}$$

Example 5.4

Find the frequency and periodic time of a sinusoidal voltage having a peak value 9 V and an instantaneous value of 4.9 V, 0.02 seconds after passing through zero.

Using the general equation

$$4.9 = 9 \sin 2\pi \times f \times 0.02$$

so that

$$\sin 2\pi f \times 0.02 = \frac{4.9}{9}$$

$$= 0.5444$$

From tables the angle having a sine equal to 0.5444 is 32° 59', that is, 32.98° and 32.98° = (32.98/360) × 2π rad. Therefore

$$2\pi f \times 0.02 = \frac{32.98}{360} \times 2\pi$$

(Note the right-hand side was not evaluated since 2π will cancel.) Therefore

$$f = \frac{32.98}{360 \times 0.02}$$

$$= 4.58 \text{ Hz}$$

PHASE RELATIONSHIPS BETWEEN TIME-VARYING QUANTITIES

If two quantities which change in value with time, change in exactly the same way at the same time they are said to be *in phase*. They do not necessarily have to have the *same* value at the same time to be in phase but what value they do have must be changing in the same way. For example, two sinusoidally alternating voltages or currents which rise and fall together, reach peak values together, pass through zero together, and have the same polarity as each other at the same time are in phase whether or not their peak values are equal. (With different peak values, the only time that the values of the two quantities are the same is when they pass through zero.)

To be out of phase one quantity must reach its peak value either before or after the other one does. If the first quantity reaches its peak *before* the second it is said to *lead* the second and if the first quantity reaches its peak *after* the second it is said to *lag* the second (see figure 5.6).

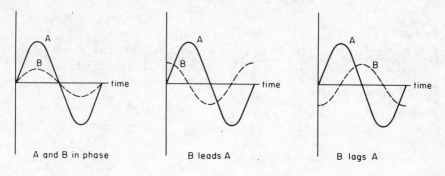

Figure 5.6

Phase relationships between sinusoidal functions

We have seen that by plotting the vertical displacement of the end of a rotating arm from the horizontal against a base of time we obtain a sinusoidal graph. Further, we have seen that the equation of the graph is

   instantaneous value = maximum value × sin (2π × frequency × time)

For an electric current, for example

$$i = I_{pk} \sin 2\pi f t$$

or

$$i = I_{pk} \sin \omega t$$

where

i = instantaneous value

I_{pk} = maximum or peak value

f = frequency

t = time

ω = angular velocity of arm in radians/second

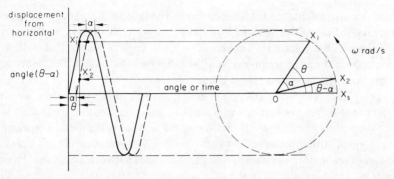

Figure 5.7

Now let us consider two such sinusoidal waves displaced in phase. In figure 5.7 two arms, OX_1 and OX_2, are shown on the right-hand side of the figure. Suppose these arms are fixed at an angle α to each other and are able to rotate about O. The sine wave shown on the left as a full line results from plotting the vertical distance of X_1 from the horizontal, as OX_1 rotates, against either a base of the angle between OX_1 and the horizontal axis or a base of time, the time starting as OX_1 passes through OX_s. If the vertical displacement of X_2 from the horizontal as OX_2 rotates is plotted to the same base the dotted sine wave results.

As shown, the dotted sine wave lags the full-line sine wave by an angle α, for when the dotted wave passes through zero, that is, when OX_2 lies along OX_s, the arm OX_1 has advanced by an angle α. Thus when the angle between OX_1 and the horizontal is θ, for example, the angle between OX_2 and the horizontal is $\theta - \alpha$. At this point X_1 is projected across to X_1' and X_2 to X_2' on the diagram. The value of the displacement of X_1 from the horizontal is $OX_1 \sin \theta$ and of X_2 from the horizontal is $OX_2 \sin (\theta - \alpha)$ and since $\theta = \omega t$, as shown previously (where t is time in seconds measured from the instant when the arm OX_1 lies along OX_s) then

instantaneous value of displacement of $X_1 = OX_1 \sin \omega t$

instantaneous value of displacement of $X_2 = OX_2 \sin (\omega t - \alpha)$

Inserting specific values in these equations we see that when $\omega t = \pi/2$ the displacement of X_1 is OX_1, that is, a maximum, but the displacement of X_2 is $OX_2 \sin [(\pi/2) - \alpha]$. The displacement of X_2 is not a maximum until $\omega t - \alpha = \pi/2$, that is $\omega t = [(\pi/2) + \alpha]$ at which time the displacement of X_1 is $OX_1 \sin [(\pi/2) + \alpha]$. At this point OX_1 has moved into the second quadrant and the displacement of X_1 has passed its peak and is falling in value again.

In general the appearance of an additional angle to ωt indicates a lag or lead so that, for example, if

$$v_1 = V_{1pk} \sin \omega t \text{ and } v_2 = V_{2pk} \sin (\omega t - \alpha)$$

where v_1, v_2 are instantaneous values of, say, a voltage and V_{1pk}, V_{2pk} are maximum values of that voltage then v_2 lags v_1 by an angle α. If the sign were positive (to give $\omega t + \alpha$) the voltage v_2 would *lead* v_1 by an angle α. Notice that both waveforms have the same frequency and the same periodic time. If the frequency is not the same then the phase angle is not constant.

So far we have referred to OX_1, OX_2, etc. as 'arms'. In fact since they are used to project time-varying waveforms which have a phase relationship with one another such arms are commonly called *phasors* and are often used to represent time-varying quantities instead of using the sinusoidal waveforms. Angles between the phasors then denote the phase difference (in time) between quantities. Phasor diagrams are considered again in the next chapter.

Example 5.5

State the phase relationships between the two quantities in each of the following pairs.

(a) $v_1 = 100 \sin \theta$; $v_2 = 25 \sin [\theta - (\pi/3)]$

(b) $i_1 = 5 \sin \theta$; $i_2 = 4 \sin [\theta + (\pi/6)]$

(c) $v_1 = 50 \sin \omega t$; $v_2 = 25 \sin [\omega t + (\pi/2)]$

(d) $v_1 = 10 \sin \omega t$; $v_2 = 12 \sin [\omega t - (\pi/12)]$

(a) v_2 lags v_1 by $\pi/3$ rad ($60°$)

(b) i_2 leads i_1 by $\pi/6$ rad ($30°$)

(c) v_2 leads v_1 by $\pi/2$ rad ($90°$)

(d) v_2 lags v_1 by $\pi/12$ rad ($15°$)

Example 5.6

Determine the maximum value, r.m.s. value and instantaneous value at $t = 5$ ms of the following voltages: $v_1 = 25 \sin 314.2t$ and $v_2 = 15 \sin [314.2t + (\pi/3)]$. What is the phase relationship between these voltages?

For v_1, maximum value is 25 V; r.m.s. value is 0.707×25, that is, 17.68 V. When $t = 5$ ms

$$v_1 = 25 \sin 314.2 \times 5 \times 10^{-3}$$

$$= 25 \sin 1.571$$

and since 1.571 rad. = $90°$

$$v_1 = 25 \text{ V}$$

For v_2, maximum value is 15 V; r.m.s. value is 0.707 × 15, that is, 10.61 V. When t = 5 ms

$$v_2 = 15 \sin [1.571 + (\pi/3)]$$

$$= 15 \sin 2.62$$

and since 2.62 rad = 150°, sin 2.62 = sin 150° = sin 30° = 0.5 and

$$v_2 = 15 \times 0.5 = 7.5 \text{ V}$$

The phase relationship is that v_2 *leads* v_1 by $\pi/3$ rad (60°).

Example 5.7

An alternating current i amperes at any time t seconds is given by $i = 20 \sin (100\pi t + 0.2)$ the angle being in radians. Find (a) the values of i when $t = 0$ and when $t = 10$ ms,   (b) one value of t at which $i = 0$ and one at which $i = 20$ A.

(a) When $t = 0$, $i = 20 \sin 0.2$. 0.2 rad = 11.46° and sin 11.46° = 0.1987, thus

$$i = 20 \times 0.1987$$

$$= 3.974 \text{ A at } t = 0$$

When $t = 10$ ms, that is, 0.010 seconds

$$i = 20 \sin (100\pi \times 0.01 + 0.2)$$

$$= 20 \sin (\pi + 0.2)$$

Now $\pi + 0.2$ rad = 191.46° and sin 191.46° = − sin 11.46° = − 0.1987, so that

$$i = - 20 \times 0.1987$$

$$= - 3.974 \text{ A at } t = 0.01 \text{ seconds}$$

(b) When $i = 0$, $20 \sin (100\pi t + 0.2) = 0$ so that sin $(100\pi t + 0.2) = 0$ and $100\pi t + 0.2 = 0$, or π, or 2π, etc. (if the sine is zero the angle could be 0, π, 2π, 3π, etc.). If we take 0 as the right-hand side the time is negative. This implies that the current passes through zero before $t = 0$, that is, before timing commences on a graph showing the waveform of the current; so that if

$$100\pi t + 0.2 = 0$$

$$t = \frac{- 0.2}{100\pi}$$

$$= - 0.637 \text{ ms}$$

and if

$$100\pi t + 0.2 = \pi$$

then

$$t = \frac{\pi - 0.2}{100\pi}$$

$$= 0.00937 \text{ seconds}$$

$$= 9.37 \text{ ms}$$

which is the first positive answer for t after t is zero.
   When $i = 20$ A

$$20 \sin (100\pi t + 0.2) = 20$$

and

$$\sin (100\pi t + 0.2) = 1$$

so that

$$100\pi t + 0.2 = \frac{\pi}{2} \text{ or } \frac{5\pi}{2} \text{ or } \frac{9\pi}{2}, \text{ etc.}$$

If

$$100\pi t + 0.2 = \frac{\pi}{2}$$

$$100\pi t = \frac{\pi}{2} - 0.2$$

$$t = (\frac{\pi}{2} - 0.2) \div 100\pi$$

$$= 0.0044 \text{ seconds}$$

$$= 4.4 \text{ ms}$$

This again is the first value of t after $t = 0$ at which $i = 20$ A

Example 5.8

Two alternating voltages are given by

$$v_1 = 100 \sin 50\pi t$$

$$v_2 = 80 \sin (50\pi t - \pi/3)$$

where t is in seconds and the angle is in radians. Determine the periodic time and frequency of each voltage. Also determine the phase difference in seconds between the voltages.

For v_1: $2\pi f = 50\pi$, where f is frequency, then $f = 25$ Hz and

$$\text{periodic time} = \frac{1}{25} \text{ seconds}$$

$$= 0.04 \text{ s}$$

$$= 40 \text{ ms}$$

For v_2: $2\pi f = 50\pi$ and $f = 25$ Hz, then

$$\text{periodic time} = 40 \text{ ms as before}$$

The phase angle is $\pi/3$ rad with v_2 lagging v_1 and the periodic time is 40 ms, therefore 2π rad (360°) corresponds to 40 ms; $\pi/3$ rad (60°) corresponds to 40/6 that is, 6.67 ms. The phase difference is thus 6.67 ms.

Example 5.9

Two currents have instantaneous values given by $i_1 = 5 \sin 50\pi t$ amperes and $i_2 = 3 \sin [50\pi t - (\pi/4)]$ amperes, where t is time in seconds. Plot the graphs of current against time for each current between $t = 0$ and $t = 20 \times 10^{-3}$ s. By adding the instantaneous values of the currents at various times determine the resultant current waveform and find the equation for its instantaneous value in terms of t (see figure 5.8).

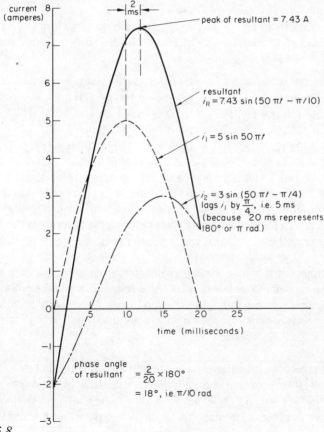

Figure 5.8

Exercise 5.2

Where relevant in the following questions sinusoidal variation may be assumed.

(1) Determine the peak value, r.m.s. value, frequency and periodic time of a current $i = 3.4 \sin 200\pi t$ where t is in seconds and the angle is in radians.

(2) An alternating voltage has a peak value of 25 V and a periodic time of 15 ms. Write down the equation for the instantaneous value of the voltage in terms of time in seconds. Find the instantaneous value after 9.3 ms.

(3) The voltage v and current i in a certain circuit are related to time t by the equations $v = 25 \sin 40\pi t$ (volts) and $i = 2.5 \sin [40\pi t + (\pi/3)]$ amperes. Find the instantaneous value of power in the circuit (the product of voltage $\times$ current) when $t = 5$ ms.

(4) Find the maximum value, r.m.s. value, average value over half a cycle and frequency of the following variable quantities. Each quantity is expressed as a function of time t in seconds.

(a) $v = 25 \sin 100\pi t$
(b) $i = 4 \sin 200\pi t$
(c) $e = 115 \sin [50\pi t + (\pi/3)]$
(d) $i = 25 \sin 2000t$
(e) $i = 7.9 \sin [3142t - (\pi/6)]$

(f) $e = 27.5 \sin [1000t - (\pi/2)]$
(g) $25 v = 100 \sin 200\pi t$
(h) $0.1i = 0.01 \sin [100\pi t - (\pi/5)]$
(i) $10i = 25 \cos 200\pi t$
(j) $5 v = 125 \cos [200\pi t + (\pi/3)]$

(5) Two alternating voltages having the same frequency 50 Hz and the same peak value 100 V are displaced in phase such that one voltage has an instantaneous value of 30 V (going positive) when the other has an instantaneous value of 10 V (going positive). Determine the phase angle between the two voltages and the value of the second when the first is at a positive peak.

(6) Two currents of instantaneous values denoted by i_1 and i_2 are alternating at 100 Hz. The maximum value of i_1 is 10 A and of i_2 is 5 A and the phase difference between the two is such that i_2 leads i_1 by $\pi/12$ rad. Write down the equations for i_1 and i_2 in terms of time t seconds and calculate the time at which each current reaches the value 3 A going positive.

(7) Two voltages have instantaneous values given by $v_1 = 100 \sin (100\pi t)$ and $v_2 = 50 \sin [(100\pi t + (\pi/6)]$ where t is time in seconds. Plot the graphs v_1/t and v_2/t between $t = 0$ and $t = 40$ ms. By adding the instantaneous values of the currents at various times determine the resultant voltage waveform and find the equation for its instantaneous value in terms of t.

(8) Two alternating currents of frequency 25 Hz, peak values 5 A and 3 A respectively, are displaced in phase such that the current with maximum value 5 A lags the other by $\pi/3$ rad. Plot the graph of these currents against angle in radians between 0 and 4π rad and by graphical addition find the equation of the resultant current, expressing its phase displacement in relation to the current having maximum value 5 A.

(9) The r.m.s. value of an alternating voltage is 19 V. Its periodic time is 15 ms. Determine the equation relating the instantaneous value of this voltage to time in seconds. Determine similarly the equation of a current of peak value 7 A alternating at the same frequency if its positive peak value occurs 2 ms before the positive voltage peak occurs.

TRIGONOMETRIC IDENTITIES

In *First-year Technician Mathematics* the functions sine θ, cosine θ and tangent θ were defined in terms of the sides of a right-angled triangle. The relationship $\tan \theta = \sin \theta / \cos \theta$ was shown to be true. Other relationships or *identities*, which are proved below, are

$$\sin^2\theta + \cos^2\theta = 1$$

$$\tan^2\theta + 1 = \sec^2\theta$$

$$1 + \cot^2\theta = \csc^2\theta$$

These identities, which are true for all values of θ, are useful in simplifying equations involving trigonometric functions.

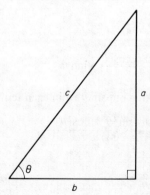

Figure 5.9

Consider figure 5.9 which shows a right-angled triangle of sides a, b and c, the hypotenuse being c. From Pythagoras' theorem

$$a^2 + b^2 = c^2$$

If we divide through by c^2 we obtain

$$\frac{a^2}{c^2} + \frac{b^2}{c^2} = 1$$

and since $a/c = \sin\theta$ and $b/c = \cos\theta$, this becomes

$$\sin^2\theta + \cos^2\theta = 1$$

Dividing through by $\cos^2\theta$

$$\frac{\sin^2\theta}{\cos^2\theta} + 1 = \frac{1}{\cos^2\theta}$$

and since $\tan\theta = \sin\theta/\cos\theta$ and $1/\cos\theta = \sec\theta$

$$\tan^2\theta + 1 = \sec^2\theta$$

Similarly, dividing $\sin^2\theta + \cos^2\theta = 1$ by $\sin^2\theta$ we have

$$1 + \frac{\cos^2\theta}{\sin^2\theta} = \frac{1}{\sin^2\theta}$$

and since $\cos^2\theta/\sin^2\theta = 1/\tan^2\theta = \cot^2\theta$ and $1/\sin^2\theta = \operatorname{cosec}^2\theta$

$$1 + \cot^2\theta = \operatorname{cosec}^2\theta$$

These relationships are called identities rather than equations because they are true for *all* values of θ rather than any particular value or number of values.

Example 5.10
Prove that

$$\frac{\sin^3\theta + \sin\theta\,\cos^2\theta}{\cos\theta} = \tan\theta$$

Considering the LHS and taking out $\sin\theta$ from each term in the numerator we have

$$\text{LHS} = \frac{\sin\theta\,(\sin^2\theta + \cos^2\theta)}{\cos\theta}$$

and since $\sin^2\theta + \cos^2\theta = 1$

$$\text{LHS} = \frac{\sin\theta}{\cos\theta}$$

$$= \tan\theta$$

$$= \text{RHS}$$

Example 5.11

Prove that

$$\frac{\sin^2\theta}{1 + \cos\theta} = 1 - \cos\theta$$

Considering the LHS, since $\sin^2\theta + \cos^2\theta = 1$ and therefore $\sin^2\theta = 1 - \cos^2\theta$, we have

$$\text{LHS} = \frac{1 - \cos^2\theta}{1 + \cos\theta}$$

The numerator may be written $(1 - \cos\theta)(1 + \cos\theta)$ by writing $(1 - \cos^2\theta)$ as the difference of two squares. Thus

$$\text{LHS} = \frac{(1 - \cos\theta)(1 + \cos\theta)}{1 + \cos\theta}$$

$$= 1 - \cos\theta$$

$$= \text{RHS}$$

Example 5.12

If $\tan\theta = 1.732$, find $\sin\theta$, $\cos\theta$, $\sec\theta$ and $\csc\theta$ *without using tables.*

Since $\sec^2\theta = \tan^2\theta + 1$, we have $\sec^2\theta = 1.732^2 + 1 = 4$

and

$$\sec\theta = 2$$

so that

$$\cos\theta = \frac{1}{\sec\theta} = 0.5$$

Since

$$\sin^2\theta = 1 - \cos^2\theta$$

$$\sin^2\theta = 1 - 0.5^2 = 0.75$$

and

$$\sin\theta = 0.866$$

thus

$$\csc\theta = \frac{1}{\sin\theta}$$

$$= \frac{1}{0.866} = 1.155$$

Exercise 5.3

Show that the following identities (1 to 11) are true for all values of θ.

(1) $\sin^4\theta - \cos^4\theta = \sin^2\theta - \cos^2\theta$

(2) $1 - \sin^4\theta = \cos^2\theta \sin^2\theta + \cos^2\theta$

(3) $\sin^2\theta + \cos^4\theta = 1 - \sin^2\theta \cos^2\theta$

(4) $\operatorname{cosec}^2\theta + \tan^2\theta = \cot^2\theta + \sec^2\theta$

(5) $\dfrac{\sin^2\theta + \cot^2\theta}{\operatorname{cosec}^2\theta - \cos^2\theta} = 1$

(6) $\dfrac{1 - \cos^2\theta}{\operatorname{cosec}^2\theta - 1} = \cos^2\theta$

(7) $\dfrac{1 + \cot^2\theta}{1 + \tan^2\theta} = \cot^2\theta$

(8) $\dfrac{(\sin\theta + \cos\theta)^2 - 2\cos\theta\sin\theta}{\operatorname{cosec}^2\theta} = \dfrac{1}{1 + \cot^2\theta}$

(9) $\operatorname{cosec}^4\theta\,(\sin^2\theta - \sin^2\theta\cos^2\theta) = 1$

(10) $(\sin\theta + \cos\theta)^2 = 2\sin\theta\cos\theta$

(11) $\cos^2\theta\tan^2\theta + \cos^2\theta\cot^2\theta = \sin^2\theta + \cos^2\theta\operatorname{cosec}^2\theta - \cos^2\theta$

(12) If $\tan\theta = 1$ find the value of $\sec^2\theta$ without using tables.

(13) If $\sin\theta = 0.25$ find the value of $\tan^2\theta\sec^2\theta$ without using tables.

(14) Find the value of $(\sec^2\theta/\cot^2\theta)^2$ without tables if $\sin\theta = 0.184$.

TRIGONOMETRIC EQUATIONS

The identities given above may be used to simplify and solve trigonometric equations.

Example 5.13

Find the values of θ between $0°$ and $360°$ that satisfy the equation
$5 - 4\cos^2\theta + 4\sin\theta = 0$.

If the identity $\cos^2\theta = 1 - \sin^2\theta$, is used, a quadratic equation in $\sin\theta$ may be obtained. Thus

$$5 - 4\,(1 - \sin^2\theta) + 4\sin\theta = 0$$

$$5 - 4 + 4\sin^2\theta + 4\sin\theta = 0$$

$$4\sin^2\theta + 4\sin\theta + 1 = 0$$

therefore

$$(2\sin\theta + 1)^2 = 0$$

and $\sin\theta = -\frac{1}{2}$ if the bracketed expression is to be zero.

The sine of an angle is negative in the third quadrant and fourth quadrant. Since the acute angle having a sine equal to $\frac{1}{2}$ is $30°$ the angles are $180 + 30$ that is, $210°$ and $360 - 30$, that is, $330°$.

Example 5.14

Solve, for θ between $\theta = 0°$ and $360°$, the equation

$$\sin^2\theta - 1.707\sin\theta\cos\theta + 0.707\cos^2\theta = 0$$

Divide through by $\cos^2\theta$ to give

$$\frac{\sin^2\theta}{\cos^2\theta} - 1.707\,\frac{\sin\theta\cos\theta}{\cos^2\theta} + 0.707 = 0$$

that is

$$\tan^2\theta - 1.707\tan\theta + 0.707 = 0$$

Factorising

$$(\tan\theta - 1)\,(\tan\theta - 0.707) = 0$$

and $\tan\theta = 1$ or 0.707 for this to be true.

The tangent is positive in the first and third quadrants and since the acute angle having a tangent equal to 1 is $45°$ and the acute angle having a tangent equal to 0.707 is $35°\ 16'$ the values of θ are $45°$, $180° + 45°$, $35°\ 16'$, $180° + 35°\ 16'$. That is, $45°$, $225°$, $35°\ 16'$, $215°\ 16'$.

Exercise 5.4

Solve the following equations for θ for values between $0°$ and $360°$.

(1) $1.5 + 1.5\sin\theta - \cos^2\theta = 0$

(2) $0.5\sin\theta - \cos^2\theta + 1.0625 = 0$

(3) $1.64 - \sin^2\theta - 1.6 \cos \theta = 0$

(4) $\cos^2\theta = 0.36 \, (\sin^2\theta + \cos^2\theta)$

(5) $\tan \theta \cos \theta + \cos^2\theta + 0.4 = 0 \div 0$

(6) $\sec^2\theta + 3 \sec \theta = 4$

(7) $\tan^2\theta + 3 \sec \theta - 2.7 = 0$

(8) $\operatorname{cosec}^2\theta - \cot \theta - 4 = 0$

(9) $\tan^2\theta - \sec \theta - 2 = 0$

(10) $(\tan^2\theta + 1)(\cot^2\theta + 1) = 30$

SOLUTION OF TRIANGLES

In *First-year Technician Mathematics* we examined various problems involving triangles and methods for determining angles and side lengths using the basic trigonometric functions. There are many different types of problem of this nature in engineering. One type of particular importance is concerned with vector and phasor diagrams which will be discussed in the next chapter. This chapter introduces a further two rules useful for solving triangles. These are the sine rule and the cosine rule and they are used in the solution of triangles that are not right-angled.

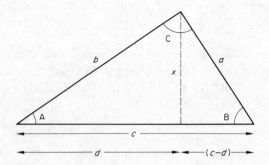

Figure 5.10

The sine rule

In the triangle shown in figure 5.10

$$\sin A = \frac{x}{b} \text{ and } \sin B = \frac{x}{a}$$

where x is the length of the perpendicular to base c drawn from the opposite vertex. Thus

$$x = b \sin A = a \sin B$$

or

$$\frac{a}{\sin A} = \frac{b}{\sin B}$$

Similarly, by drawing a perpendicular to side a produced from the opposite vertex, we could show that

$$c \sin B = b \sin C$$

so that, in general

$$\frac{a}{\sin A} = \frac{b}{\sin B} = \frac{c}{\sin C}$$

which is true, for all triangles whether the angles concerned are acute or obtuse. This rule is called *the sine rule*.

The cosine rule

Consider again the triangle shown in figure 5.10. Applying Pythagoras' theorem to each right-angled triangle in turn we have

$$x^2 + d^2 = b^2 \qquad \text{and } x^2 + (c - d)^2 = a^2$$

so that

$$x^2 = b^2 - d^2 \qquad \text{and } x^2 = a^2 - (c - d)^2$$

Therefore

$$b^2 - d^2 = a^2 - (c - d)^2$$

and

$$a^2 = b^2 - d^2 + (c - d)^2$$
$$= b^2 - d^2 + c^2 - 2cd + d^2$$
$$= b^2 + c^2 - 2cd$$

Now from the left-hand triangle

$$\cos A = \frac{d}{b}$$

that is

$$b \cos A = d$$

and substituting this into the equation for a^2 we obtain

$$a^2 = b^2 + c^2 - 2 bc \ \cos A$$

which is *the cosine rule.* Similarly it may be shown that

$$b^2 = a^2 + c^2 - 2ac \cos B$$

and

$$c^2 = a^2 + b^2 - 2ab \cos C$$

The cosine rule is used in the solution of triangles when two sides and the included angle are known or when three sides are known; otherwise the sine rule may be used.

Example 5.15

A vertical mast is supported by two ropes anchored at ground level so that they and the mast are in the same plane. If the lengths of the ropes are 75 m and 60 m and they are anchored 120 m from each other find the height of the mast.

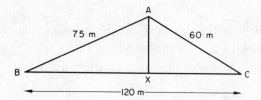

Figure 5.11

The diagram is shown in figure 5.11 in which AB and AC represent the ropes and AX the mast. Now

$$\sin A\hat{B}X = \frac{AX}{AB} \quad \text{or} \quad \sin A\hat{C}X = \frac{AX}{AC}$$

and AB (or AC) is known. Thus we need to find either $A\hat{B}X$ (or $A\hat{C}X$) and since three sides of the triangle are known the cosine rule is used. Applying the cosine rule to △ABC to find $A\hat{B}X$

$$AC^2 = AB^2 + BC^2 - 2\ AB\ BC\ \cos\ A\hat{B}X$$

Thus

$$2\ AB\ BC\ \cos\ A\hat{B}X = -\ AC^2 + AB^2 + BC^2$$

and

$$\cos\ A\hat{B}X = \frac{-AC^2 + AB^2 + BC^2}{2A\ B\ BC}$$

$$= \frac{-\ 60^2 + 75^2 + 120^2}{2 \times 75 \times 120}$$

$$= 0.9125$$

Thus $A\hat{B}X = 24°\ 9'$ from tables and $\sin\ A\hat{B}X = 0.4090$ from tables. Therefore from above

$$AX = AB\ \sin\ A\hat{B}X$$

$$= 75 \times 0.4090$$

$$= 30.68\ m$$

Example 5.16

A ship steaming due east at 65 km/h is sighted by a stationary observer situated at a bearing S 25° E from the ship. One hour later the ship is at a bearing N 35° E from the observer. Determine the distance between the observer and the ship at both sightings and also the shortest distance between observer and ship as the ship proceeds.

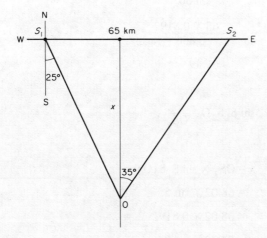

Figure 5.12

The diagram showing ship positions S_1, S_2 and observer O is shown in figure 5.12. Note the given angles on the diagram, O being S 25° E from S_1 and S_2 being N 35° E from O. Since the time between sightings is one hour, the length $S_1 S_2$ is 65 km. Length x is the shortest distance between ship and observer and lengths OS_1 and OS_2 are also required.

Examination of the diagram shows that in $\triangle S_1 O S_2$, angle S_1 is $90 - 25$ that is, 65° and angle S_2 is $90 - 35$ that is, 55°. Since two angles and a side are known in $\triangle S_1 O S_2$ the sine rule may be used. Thus

$$\frac{OS_2}{\sin O S_1 S_2} = \frac{OS_1}{\sin S_1 S_2 O} = \frac{S_1 S_2}{\sin S_1 O S_2}$$

Now $O\hat{S}_1 S_2 = 65°$, $S_1 \hat{S}_2 O = 55°$; $S_1 S_2 = 65$ km and thus $S_1 \hat{O} S_2 = 180 - 65 - 55 = 60°$. Therefore

$$\frac{OS_2}{\sin 65} = \frac{OS_1}{\sin 55} = \frac{65}{\sin 60}$$

and

$$OS_2 = \frac{65 \sin 65}{\sin 60}$$

$$= \frac{65 \times 0.9063}{0.866}$$

$$= 68.02$$

and

$$OS_1 = \frac{65 \sin 55}{\sin 60}$$

$$= \frac{65 \times 0.8192}{0.866}$$

$$= 61.49$$

Now

$$\sin S_1 \hat{S}_2 O = \frac{x}{OS_2}$$

so that

$$x = OS_2 \times \sin S_1 \hat{S}_2 O$$

$$= 68.02 \times \sin 55°$$

$$= 68.02 \times 0.8192$$

$$= 55.72 \text{ km}$$

Example 5.17

A triangle ABC has AB = 19 cm, AC = 16 cm and AB̂C = 50°. Find the remaining sides and angles.

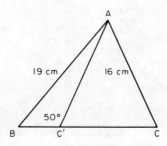

Figure 5.13

As shown in figure 5.13 a triangle with dimensions such as those given may have more than one shape, that given as ABC and that given as ABC'. C' is a point on BC and both C and C' are 16 cm from A, but AB (at 19 cm) and AB̂C (at 50°) remain unchanged. Only one method of solution is needed, the two possible values of the angles BÂC and AĈB (that is, BÂC and BÂC', AĈB and AĈ'B) being provided by the fact that any positive sine value results from an angle in the first quadrant and an angle in the second quadrant.

$$\frac{AB}{\sin ACB} = \frac{AC}{\sin ABC} = \frac{BC}{\sin BAC}$$

that is

$$\frac{19}{\sin ACB} = \frac{16}{\sin 50} = \frac{BC}{\sin BAC}$$

so that

$$\sin A\hat{C}B = \frac{19 \sin 50}{16}$$

$$= \frac{19 \times 0.766}{16}$$

$$= 0.91$$

thus AĈB = 65° 30′ or 180° − 65° 30′ that is, 114° 30′. Therefore BÂC = 180° − 50° − 65° 30′ or 180° − 50° − 114° 30′ (since the sum of angles = 180°) that is, 64° 30′ or 15° 30′, and since

$$BC = \frac{AC}{\sin ABC} \times \sin BAC$$

$$BC = \frac{16 \sin 64° \; 30'}{\sin 50} \; \text{or} \; \frac{16 \times \sin 15° \; 30'}{\sin 50}$$

$$= \frac{16 \times 0.9026}{0.766} \; \text{or} \; \frac{16 \times 0.2672}{0.766}$$

$$= 18.85 \qquad \text{or} \; 5.58$$

The complete answer is therefore

$$BC = 18.85 \qquad B\hat{A}C = 64° \; 30' \qquad A\hat{C}B = 65° \; 30'$$

$$BC = \; 5.58 \qquad B\hat{A}C = 15° \; 30' \qquad A\hat{C}B = 114° \; 30'$$

$AB = 19$, $AC = 16$ and $A\hat{B}C = 50°$ for both triangles as given.)

The possibility of ambiguity, that is, two possible answers, should be investigated in all problems unless it is definitely known not to exist.

Areas of triangles

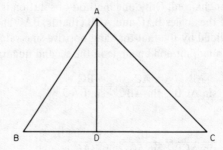

Figure 5.14

As given in earlier chapters the area of any triangle is equal to one half of the product base × height. So that for $\triangle ABC$, (figure 5.14) the area of the triangle is $\frac{1}{2} \times BC \times AD$. Denoting AB, BC and CA by c, a and b (being the notation used earlier) we have

$$\text{Area} = \frac{1}{2} \times a \times AD$$

and since

$$\sin A\hat{C}D = \frac{AD}{AC} = \frac{AD}{b}$$

then $AD = b \sin C$ (again using the previous notation) so that

$$\text{Area of } \triangle ABC = \frac{1}{2} ab \sin \hat{C}$$

similarly
$$= \frac{1}{2} bc \sin \hat{A}$$

and
$$= \frac{1}{2} ac \sin \hat{B}$$

This general formula is most useful when two sides of the triangle and an included angle are given.

It may also be shown that if s is half the sum of the sides, that is $2s = a + b + c$, then the area of the triangle is

$$\sqrt{[s(s-a)(s-b)(s-c)]}$$

This equation may be used, therefore, when all three sides of a triangle are given.

Example 5.18

Find all three sides of triangle ABC in which $a = 4.6$, $b = 5.9$ and $\hat{C} = 35°$. Determine the area of the triangle using both methods given above.

To find c, apply the cosine rule

$$c^2 = 4.6^2 + 5.9^2 - 2 \times 4.6 \times 5.9 \times \cos 35$$

$$= 21.16 + 34.81 - 44.47$$

$$= 11.50$$

hence

$$c = 3.391$$

Therefore

$$2s \text{ (sum of sides)} = 5.9 + 4.6 + 3.391$$

$$= 13.89$$

so that $s = 6.95$ and

$$\text{area} = \sqrt{[6.95(6.95 - 5.9)(6.95 - 4.6)(6.95 - 3.391)]}$$

$$= \sqrt{(6.95 \times 1.05 \times 2.35 \times 3.56)}$$

$$= \sqrt{61.05}$$

$$= 7.8$$

But

$$\text{Area of triangle} = \frac{1}{2} ab \sin \hat{C}$$

$$= \frac{1}{2} \times 4.6 \times 5.9 \times \sin 35$$

$$= 7.8 \text{ confirming the previous answer}$$

Exercise 5.5

(1) A triangle ABC has the following dimensions (using the notation already given). Find the remaining sides and angles.

(a) $a = 10, b = 15, \hat{C} = 35°$

(b) $b = 4, c = 7, \hat{A} = 28°$

(c) $a = 12, b = 17, \hat{C} = 10°$

(d) $a = 10, c = 5, \hat{B} = 35°$

(e) $a = 15, b = 20, \hat{B} = 80°$

(f) $c = 25, a = 30, \hat{A} = 17°$

(g) $a = 13, b = 26, \hat{A} = 15°$

(h) $a = 18.5, c = 14.5, \hat{A} = 52°$

(i) $a = b = 15, \hat{B} = 13°$

(2) A mast is supported by two ropes, the mast and ropes all being in the same plane. If the ropes are 10 m and 12 m long and the angle that the 10 m rope makes with the ground is 30°, find the distance between the ground anchorage points of the ropes. Find also the height of the mast assuming that both ropes are tied at the mast top.

(3) A car moving due east at 50 km/h is sighted by an observer first at a bearing N 35° W from the observer and second at a bearing N 22° E from the observer. If the time between sightings is 45 minutes, calculate the distance between the observer and the car on each occasion of sighting.

(4) Two sides of a triangle are 7 cm and 4 cm and the included angle is 42°. Calculate the area of the triangle.

(5) An airport A is 500 km NW of a radar beacon B. Another radar beacon C is 300 km from A and lies due south of A. Find the bearing of C from B and the distance between C and B.

(6) A tower AB 30 m high is observed from two points C and D such that directions CB and DB are at right angles, B being the base of the tower. If the angle of elevation of the top of the tower A from C is 40° and from D is 50° find the distance between C and D.

(7) A parallelogram of sides 20 cm and 10 cm has internal acute angles each equal to 25°. Find the length of the diagonal.

(8) Using the sine rule show that for any triangle ABC, $a^2 \sin^2 B = b^2 (1 - \cos^2 A)$ where a and b are side lengths opposite angles A and B respectively.

(9) Find the largest angle in a triangle having sides of 1.2 m, 2.4 m, 2.7 m.

(10) Find the smallest angle in the triangle described in question 10.

(11) Find the area of a triangle ABC having $\hat{A} = 20°$, $\hat{B} = 40°$, $b = 3$ cm.

(12) A metal plate ABCD has dimensions BC = 3.5 cm, AC = 4.2 cm and angles $A\hat{C}D = 25°$, $D\hat{A}C = 45°$, $B\hat{C}A = 30°$. Find the area of the plate.

(Other problems requiring the use of the sine or cosine rules are at the end of chapter 6.)

6 Vectors and Phasors

In engineering and mathematics, physical quantities are those that can be measured in one way or another and the measurement used to compare one quantity with another like quantity. For example, mass is a physical quantity and the amount of matter in one mass can be measured and compared with the amount of matter in another mass. Thus we can say that a 10 kg mass, for example, has twice the amount of matter of a 5 kg mass. Similarly time is a physical quantity and a period of time of, say, 3 hours is three times as long as a period of time of 1 hour. Other physical quantities include length, velocity, acceleration, force, work, energy, power, voltage, electric current, and so on. With many physical quantities the only measurable characteristic is their magnitude or size. Such quantities, for example, mass, length, area, etc., are called *scalar* quantities. Other quantities have associated with them not only magnitude but also a physical direction in space. These quantities, which include force and velocity, are called *vector* quantities. A third group of quantities contains those that not only have magnitude but are also time varying, that is, their magnitude varies regularly or otherwise with time. These quantities—which include alternating voltages, currents and the electric and magnetic fields set up by such alternating voltages and currents—are called *phasor* quantities. In determining the sum or difference of two quantities or when performing other arithmetical processes such as multiplication or division care must be taken to first establish whether the quantities are scalar, vector, or phasor.

If one considers two masses of, say, 10 kg and 6 kg, the single mass that is equivalent to the two put together is 10 + 6 = 16 kg; in other words the two masses of 10 kg and 6 kg could be replaced by one mass of 16 kg. Mass is a scalar quantity and the straightforward addition shown is scalar addition. Similarly for other arithmetical processes, the process is applied to the numbers representing the magnitude of the quantity.

In finding the equivalent of two vector quantities or two phasor quantities, or in performing other mathematical processes, the method is less straightforward, since with vectors, the direction in which the vector acts must be considered, and with phasors, the time-varying characteristic or phase of the phasor must be considered. For example, consider two forces (vectors) of magnitudes 10 N and 6 N respectively. Scalar addition of these quantities would give a force of 10 + 6, = 16 N, but this force is not necessarily the equivalent of the original two forces;

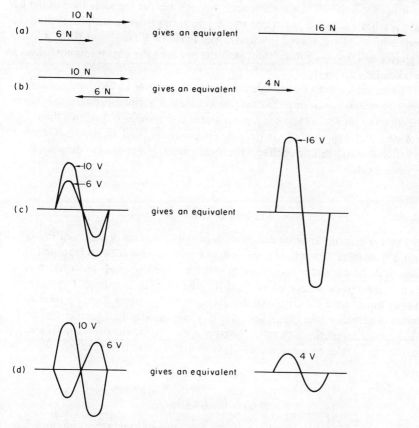

Figure 6.1

in other words, the two forces of 10 N and 6 N cannot necessarily be replaced by a single force of 16 N (see figure 6.1a). The value of the equivalent is determined by the directions in which the 10 N and 6 N forces are acting. Consider, for example, the 10 N force acting in one direction and the 6 N force acting in exactly the opposite direction. The single equivalent force is then 10 − 6, that is, 4 N acting in the direction of the original 10 N force. The only occasion when the equivalent force is 16 N is when both of the original forces act in exactly the same direction, as in figure 6.1a. Vector addition (or subtraction, etc.) must always take into account the direction of action of the vectors.

A similar situation occurs with phasor quantities. Consider two sinusoidal voltages, one of peak value 10 V, the other of peak value 6 V. If both voltages are varying in phase with each other—that is, rising to a peak value simultaneously—the single equivalent voltage is a sinusoid of peak value 16 V (figure 6.1c). If on the other hand the 10 V wave reaches a positive peak at the same time as the 6 V wave reaches a negative peak (that is, the two voltages are in antiphase) then the single equivalent voltage is a sinusoid of peak value 4 V (see figure 6.1d). Clearly in phasor addition (or subtraction, etc.) the phase of the quantities must always be taken into account.

The examples given cite simple cases of vectors acting in the same or exactly opposite directions, or of phasors acting in phase or antiphase, but usually the directions or phases are less convenient and vector or phasor diagrams then provide a useful method of mathematically processing such quantities. Geometry and trigonometry can be applied with equal ease to either kind of diagram as explained below.

VECTOR DIAGRAMS

In a vector diagram lines are drawn to show the directions of action of the various vector quantities. These lines may be drawn to a suitable scale to represent the magnitude of the vector quantities; in which case geometrical construction can then be used to determine equivalents (or whatever else is required) and the answer can be read directly from the diagram. Alternatively the diagram may be drawn merely as a guide, with lines not to scale, and standard geometrical or trigonometrical methods can then be employed to solve the problem. In a vector diagram the angles between vectors show *actual directions in space*.

PHASOR DIAGRAMS

If diagrams similar to vector diagrams are to be drawn to represent phasors, some method must be employed to show the time or phase relationship between quantities. One method of doing this has already been discussed in earlier chapters and in volume 1 of this series. In this method $360°$, or 2π radians, represents one cycle of the varying quantity, and if the peak value of one alternating quantity occurs one quarter of a cycle earlier or later than the peak value of another alternating quantity, we say that the first quantity leads or lags the second by $\frac{1}{4}$ of $360°$ (or $\frac{1}{4}$ of 2π radians), that is $90°$ (or $\pi/2$ radians). Representing the first quantity by, say $V_1 \sin \omega t$, the second quantity would then be $V_2 \sin [\omega t + (\pi/2)]$ or $V_2 \sin [\omega t - (\pi/2)]$ where V_1, V_2 indicate the maximum values of these quantities and the + or − sign in the bracket denotes lead or lag respectively (as was considered in detail in chapter 5).

Thus diagrams showing phasors can be drawn in much the same way as diagrams showing vectors, with the important difference that, whereas in vector diagrams angles denote direction in space, in phasor diagrams angles denote a difference in phase between quantities. The phase difference in seconds may be easily computed from the phase angle since $360°$ represent 1 cycle or T seconds (where T is the periodic time) so that $\theta°$ represent $\theta/360$ of 1 cycle or $\theta T/360$ seconds or if θ is the angle in radians θ rad represent $\theta/2\pi$ of 1 cycle or $\theta T/2\pi$ seconds.

The length of the lines in phasor diagrams may be drawn to represent instantaneous values, r.m.s. values or peak (maximum) values of the phasor quantities. The same kind of value should be used throughout any one diagram, that is, all r.m.s., or all peak values, etc. As with vector diagrams, phasor diagrams may be drawn to scale and the phasors may be added or subtracted by geometrical construction; or standard methods of geometry or trigonometry can be employed when the phasors are not drawn to scale. These methods will be considered after the basic examples below.

Example 6.1

Draw vector diagrams to show the following quantities and their respective directions.

(a) A force of 150 N acting horizontally along a smooth surface on a mass of 15 kg which rests on the surface. Show all forces acting on the mass.

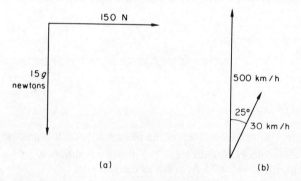

Figure 6.2

See figure 6.2a. Two forces are involved, one of 150 N acting horizontally and the weight of the 15 kg mass acting vertically downwards. This weight is equal to 15 g newtons where g is the acceleration due to gravity in metres/second² *.

*1 newton is that force which gives a mass of 1 kg an acceleration of 1 m/s². The weight of a mass of 1 kg gives the mass an acceleration of g m/s²; thus the weight of a 15 kg mass is 15 g newtons.

(b) An aeroplane is flying due north with a velocity of 500 km/h. A wind is blowing in a direction N 25° E with a velocity of 30 km/h. Show the velocities of the wind and of the aeroplane.

The diagram is shown in figure 6.2b.

Example 6.2

Draw phasor diagrams to show (a) two voltages, one being represented by the equation $v_1 = 100 \sin \omega t$, the other by the equation $v_2 = 150 \sin [\omega t + (\pi/3)]$ where v_1 and v_2 are the instantaneous values of the voltages; (b) two currents, one having an r.m.s. value of 15 A, the other having an r.m.s. value of 20 A; both currents varying sinusoidally at the same frequency with the second lagging the first by $2\pi/3$ radians.

(a) The peak values of the voltages are 100 V and 150 V with the second voltage leading the first by $\pi/3$ rad (60°). The diagram is shown in figure 6.3a. Any position may be chosen for the reference voltage, v_1, provided the other has the correct phase angle relative to it (though it is customary for the reference phasor to take the horizontal position).

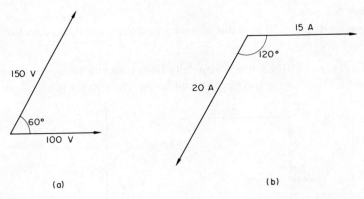

(a)   (b)

Figure 6.3

(b) Here, r.m.s. values are shown on the phasor diagram but peak values (being $\sqrt{2} \times$ r.m.s.) may equally be indicated. The diagram is shown in figure 6.3b with the first current, r.m.s. value 15 A, taken as reference.

Example 6.3

A body of mass M kilogrammes is supported by two wires, one at 60° to the vertical, the other at 45° to the vertical. The tensions in the wires are T_1 newtons and T_2 newtons respectively. Draw a vector diagram showing all forces acting on the body.

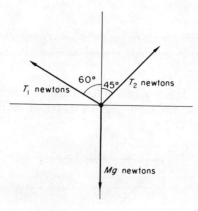

Figure 6.4

The diagram is shown in figure 6.4. Note that the weight of the body, being *Mg* newtons, acts vertically downwards.

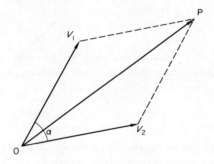

Figure 6.5

SUMMATION OF VECTORS AND PHASORS USING PARALLELOGRAMS

Two vectors (or phasors) may be summed to determine the equivalent vector (or phasor) by completion of the parallelogram two sides of which are formed by the vectors (or phasors) being summed. Consider figure 6.5 showing two vectors (or phasors) OV_1, OV_2 acting at an angle α to each other. Completing the parallelogram by drawing two more sides, one parallel to one vector*, and the other side parallel to the other vector intersecting to give point P, then O*P* is the resultant single vector that replaces the original two completely. The length of O*P* and its direction relative to either OV_1 or OV_2 may be found either by drawing the diagram to scale or by using standard geometrical or trigonometrical means. If α is a right angle the means employed could be Pythagoras' theorem, if α is not a right angle the cosine or sine rule could be employed.

*From this point on the words 'or phasor' will be omitted. All the theory applying to vector-diagram construction and solution given here applies equally to phasors unless otherwise stated.

Example 6.4

Determine the resultant force acting on the mass in example 6.1a (a) by
construction and   (b) by calculation using geometry or trigonometry. Assume
$g = 9.81$ m/s².

The diagram of figure 6.2a is repeated in figure 6.6. In this case the angle between
the two vectors to be summed is 90° and the magnitude of the vectors is 150 N
and 15 × 9.81, that is, 147.2 N.

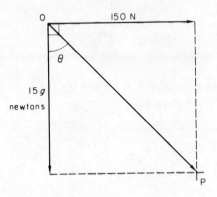

Figure 6.6

(a) Choose a suitable scale depending on the size of diagram required, say
1 cm = 20 N, and construct a vertical line of length 7.5 cm to represent 150 N.
Construct a horizontal line of length 7.36 cm as shown to represent 147.2 N.
Complete the parallelogram and join *OP* as in the diagram. The length should be
10.51 cm and since 1 cm = 20 N the magnitude of the resultant is 210.2 N. The
value of the angle θ between the resultant and the vertical by measurement is
45° 34′. Thus the resultant force acting on the body is 210.2 N acting downwards at an angle 45° 34′ to the vertical.

(b) Using the diagram in figure 6.6, the magnitude of *OP* by Pythagoras'
theorem is

$$\sqrt{(7.5^2 + 7.36^2)} \text{ cm}$$

$$= \sqrt{110.42}$$

$$= 10.51 \text{ as before}$$

The tangent of angle θ between the resultant and the vertical is 150/147.2 = 1.02
and, from tables, since tan θ = 1.02

$$\theta = 45° \ 34′ \text{ as before}$$

Note that in practice the values obtained by measurement will not in fact be as close to the calculated values as the results of this example might suggest.

Example 6.5

Obtain the resultant velocity (magnitude and direction) of the aeroplane described in example 6.1b if the wind direction is due east.

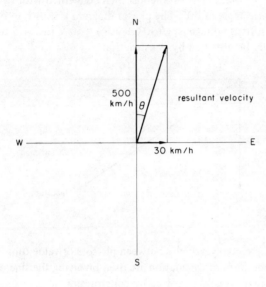

Figure 6.7

The diagram is shown in figure 6.7. This may be drawn to scale and measurements taken as required, or a calculation using the diagram may be made as follows.

$$\text{Resultant velocity} = \sqrt{(500^2 + 30^2)} \text{ km/h}$$

$$= 500.9 \text{ km/h}$$

Resultant direction is N θ° E where

$$\tan \theta = \frac{30}{500} = 0.06$$

hence

$$\theta = 3^\circ \; 26'$$

The aeroplane has a resultant velocity of 500.9 km/h in a direction N 3° 26′ E.

The examples given so far have been concerned with vectors. The methods apply equally to phasors that have a 90° phase angle between them. This is shown in the next example.

Example 6.6

In a simple series circuit consisting of a capacitor and resistor the voltage across the capacitor is 30 V and the voltage across the resistor is 40 V. Calculate the supply voltage.

In such a circuit the capacitor voltage lags the current by 90°, the resistor voltage is in phase with the current. The problem is then concerned with two phasors separated by a phase angle of 90°. The phasor diagram is shown in figure 6.8. By calculation the resultant voltage is $\sqrt{(30^2 + 40^2)} = 50$ V (a 3:4:5 triangle). This result could equally be obtained by construction.

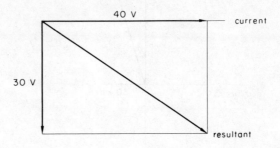

Figure 6.8

The next example involves angles between phasors of value other than 90°. Trigonometrical methods of calculation are used involving the sine or cosine rules. All the problems may equally be solved by construction.

Example 6.7

Find the resultant of (a) the two voltages and   (b) the two currents described in example 6.2.

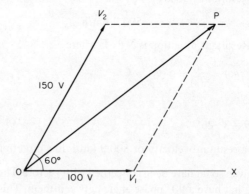

Figure 6.9

(a) The voltages are represented by $v_1 = 100 \sin \omega t$ and $v_2 = 150 \sin (\omega t + (\pi/3)]$ where v_1 and v_2 are instantaneous values. The phasor diagram is shown in figure 6.3a and is redrawn in figure 6.9 to show the resultant voltage **OP**. This may be calculated using the cosine rule as follows.

In the diagram, $\mathbf{OV_1}$ and $\mathbf{OV_2}$ represent the two voltages, and figure OV_2PV_1 is a parallelogram with $\mathbf{OV_2}$ parallel to $V_1\mathbf{P}$ and $V_2\mathbf{P}$ parallel to $\mathbf{OV_1}$. Since $V_2\hat{O}V_1 = 60°$, then the angle between $\mathbf{PV_1}$ and $\mathbf{OV_1}$ extended to X is also 60° (angle PV_1X). Thus angle $OV_1P = 120°$.

Using the cosine rule

$$OP^2 = OV_1^2 + PV_1^2 - 2 \times OV_1 \times PV_1 \times \cos OV_1P$$

and since $\mathbf{OV_1} = 100$

$$PV_1 = OV_2 = 150$$
$$O\hat{V}_1P = 120°$$

since $\cos 120° = -\cos 60° = -0.5$, then

$$OP^2 = 100^2 + 150^2 + 2 \times 100 \times 150 \times 0.5$$

$$= 47\,500$$

and

$$OP = 217.9 \text{ V (peak value of resultant voltage)}$$

The phase angle between the resultant **OP** and the reference voltage $\mathbf{OV_1}$ is angle POV_1 which may be found using the sine rule as follows. In $\triangle OPV_1$

$$\frac{PV_1}{\sin POV_1} = \frac{OP}{\sin OV_1P}$$

Thus

$$\sin POV_1 = \frac{PV_1 \times \sin OV_1P}{OP}$$

$$= \frac{150 \sin 120°}{217.9}$$

$$= 0.5961$$

Angle POV_1 is an acute angle (that is, between 0° and 90°), thus, from tables, phase angle of resultant, angle POV_1, is given by

$$P\hat{O}V_1 = 36° \; 36'$$

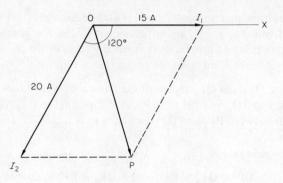

Figure 6.10

(b) The phasor diagram is shown in figure 6.3b and is repeated in figure 6.10 showing the resultant.

Applying the cosine rule to $\triangle OI_1 P$

$$OP^2 = OI_1^2 + I_1 P^2 - 2\, OI_1 \times I_1 P \times \cos OI_1 P$$

since

$$P\hat{I}_1 X = I_2 \hat{O} I_1 = 120°$$

$$O\hat{I}_1 P = 180° - P\hat{I}_1 X = 60°$$

Thus

$$OP^2 = 15^2 + 20^2 - 2 \times 15 \times 20 \cos 60$$

$$= 625 - 600 \times 0.5$$

$$= 325$$

$$OP = 18.03$$

resultant current = 18.03 A (r.m.s. value)

Phase angle $I_1 OP$ is obtained from the sine rule

$$\frac{OP}{\sin OI_1 P} = \frac{I_1 P}{\sin I_1 OP}$$

Thus

$$\sin I_1 OP = \frac{I_1 P}{OP} \sin OI_1 P$$

$$= \frac{20}{18.03} \sin 60$$

$$= 0.9607$$

and

$$I_1 \, \hat{O}P = 73° \, 52' \text{ from tables}$$

PROBLEMS INVOLVING MORE THAN TWO VECTORS

Many problems—for example, those involving a number of forces in mechanics; a series circuit containing more than two components; or a parallel circuit with more than two branches—require the determination of the resultant of more than two vectors or phasors. The procedure so far described may be used to replace two of the vectors with one resultant, then to replace the resultant and one other vector with a second resultant, and so on until all vectors are subsequently replaced by one overall resultant.

A second method, which is probably more convenient, is to use vector resolution and even when only two vectors are involved this method may be used as an alternative to the method already outlined.

Vector and phasor resolution

Resolution of vectors is the opposite process to summation of two vectors. It consists of replacing a vector by two *components* the resultant of which is the vector that is being replaced. The two components are usually chosen to lie one along the horizontal and the other along the vertical. Continuing the process for a number of vectors gives several vectors lying along the horizontal—which can then be added directly to give the resultant horizontal component—and several vectors lying along the vertical, which can then be added directly to give the resultant vertical component. The two resultant components may then be summed by Pythagoras' theorem to give the overall resultant.

To clarify this consider any vector **OP** that is the resultant of two components **OH** along the horizontal and **OV** along the vertical as shown in figure 6.11.

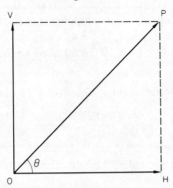

Figure 6.11

OVPH is the parallelogram from which the resultant is obtained and **PH = VO** and **VP = OH**. Let PÔH be denoted by θ.

Now sin θ = **PH/OP**, therefore **PH** = OP sin θ so that the vertical component of **OP** is given by

$$OV = PH = OP \sin \theta$$

Also cos θ = **OH/OP** and thus the horizontal component of **OP** is given by

$$OH = OP \cos \theta$$

In general, any vector or phasor **OP** at an angle θ to the horizontal may be resolved into two components given by

$$\text{horizontal component} = OP \cos \theta$$

$$\text{vertical component} \quad = OP \sin \theta$$

Example 6.8

Find the resultant of the three vectors shown in figure 6.12 and the angle it makes with the horizontal OH.

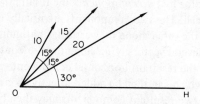

Figure 6.12

The components of the vector of magnitude 20 are 20 cos 30 along the horizontal and 20 sin 30 along the vertical. The components of the vector of magnitude 15 are 15 cos 45 along the horizontal, 15 sin 45 along the vertical. The components of the vector of magnitude 10, are 10 cos 60 along the horizontal and 10 sin 60 along the vertical.

Resultant horizontal component = 20 cos 30 + 15 cos 45 + 10 cos 60

$$= (20 \times 0.866) + (15 \times 0.707) + (10 \times 0.5)$$

$$= 32.92$$

Resultant vertical component $\quad$ = 20 sin 30 + 15 sin 45 + 10 sin 60

$$= (20 \times 0.5) + (15 \times 0.707) + (10 \times 0.866)$$

$$= 29.26$$

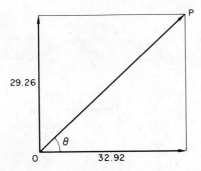

Figure 6.13

The overall resultant is shown as **OP** in figure 6.13 where θ is the required angle between resultant and horizontal, and

The resultant **OP** is given by

$$\mathbf{OP}^2 = \sqrt{(32.92^2 + 29.26^2)}$$

$$\mathbf{OP} = 44.03$$

The angle θ is obtained from

$$\tan \theta = \frac{29.26}{32.92}$$

$$= 0.8888$$

and, from tables, $\theta = 41° \, 38°$.

Example 6.9

The instantaneous values v_1, v_2 and v_3 of three voltages are given by the equations

$$v_1 = 100 \sin [\omega t + (\pi/3)]$$

$$v_2 = 50 \sin \omega t$$

$$v_3 = 75 \sin [\omega t - (3\pi/4)]$$

Determine the equation giving the instantaneous value v_R of the resultant of these three voltages.

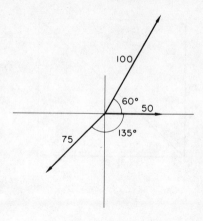

Figure 6.14

The phasor diagram is shown in figure 6.14. Voltage v_2 is taken as reference and is drawn along the horizontal. Voltage v_1 leads v_2 by $\pi/3$ rad, (that is 60°) and v_3 lags v_2 by $3\pi/4$ rad (that is, 135°).

The horizontal component of v_1 is 100 cos 60 = 50. The horizontal component of v_3 is 75 cos 135. This equals 75 × (− cos 45), which is − 53. Note that this component is negative and drawn from right to left along the horizontal in figure 6.15. Taking into account voltage v_2 which already lies along the horizontal,

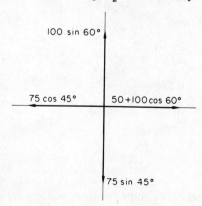

Figure 6.15

the resultant horizontal component (shown in figure 6.16) is 50 + 50 − 53 = 47. The vertical component of v_1 is 100 sin 60 = 86.6, and of v_2 is zero, since v_2 lies along the horizontal; and the vertical component of v_3 is 75 sin 135 = 75 × (− sin 45) or − 53. (This is negative and so it is drawn vertically *downwards* in figure 6.15.) The resultant vertical component is therefore 86.6 − 53 = 33.6, as shown in figure 6.16. Note that this component is positive.

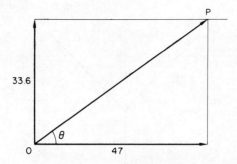

Figure 6.16

The magnitude of the resultant voltage, **OP** in figure 6.16, is given by

$$OP = \sqrt{(33.6^2 + 47^2)}$$

$$= 57.78$$

Angle θ, the phase angle between the resultant and the reference v_2, is given by

$$\tan \theta = \frac{33.6}{47} = 0.715$$

Hence

$$\theta = 35° \ 34'$$

$$= 0.6208 \ \text{rad}$$

$$= 0.1976\pi \ \text{rad}$$

(The figure 0.1976 is obtained by dividing 0.6208 by π.) Hence the equation giving the instantaneous value v_R of the resultant voltage is

$$v_R = 57.78 \sin (\omega t + 0.1976\pi)$$

or

$$v_R = 57.78 \sin (\omega t + 0.6208)$$

The phase angle may be written in terms of π or not, as required; one slight advantage of using π is that its presence indicates that the phase angle is in radians. Note from figure 6.16 that the resultant *leads* the reference voltage v_2 and the phase angle in the equation giving the answer is shown positive.

Example 6.10
A body of mass 100 kg is suspended by two strings equally displaced about a vertical line drawn through the body and each string makes an angle of 45° with the horizontal. If the system is in equilibrium calculate the tension in each string. Assume $g = 9.81 \ \text{m/s}^2$.

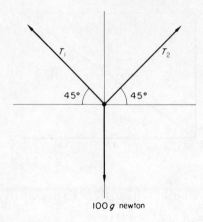

100 *g* newton

Figure 6.17

The situation is shown in figure 6.17. When any system is in equilibrium there is no resultant force acting on it. This implies that the resultant vertical component obtained by summing the vertical components of all forces present is zero. Similarly the resultant horizontal component obtained by summing the horizontal components of all forces present must also be zero if the overall resultant is zero.

Let the string tensions be T_1 N and T_2 N respectively, then the horizontal components are $-T_1 \cos 45$ and $T_2 \cos 45$. Thus

$$-T_1 \cos 45 + T_2 \cos 45 = 0$$

and

$$T_1 = T_2$$

Vertical components are $T_1 \sin 45$, $T_2 \sin 45$ and $-100g$ N. Thus

$$T_1 \sin 45 + T_2 \sin 45 - 100\,g = 0$$

and since

$$T_1 = T_2$$

$$2T_1 \sin 45 = 100g$$

that is

$$T_1 = \frac{100 \times 9.81}{2 \sin 45} \text{ N}$$

$$= 694 \text{ N}$$

The string tension is thus 694 N in each string.

This example illustrates another use of vector resolution, one in which the resultant is known and the magnitude of forces present is to be found.

Vector subtraction

To obtain the difference between two vectors the vector that has to be subtracted is drawn in the opposite direction to its normal position and then one of the methods already given is used to find the resultant of summing the remaining vector with this redrawn vector. This is illustrated in the following example.

Example 6.11

Find the vector difference between a force of 100 N acting along the horizontal and a force of 200 N acting at $120°$ to the horizontal in the second quadrant.

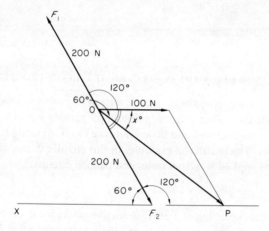

Figure 6.18

The vector diagram is shown in figure 6.18. The 200 N force is drawn in the opposite direction to its line of action, that is, along OF_2 instead of OF_1 and the resultant of the 100 N force and the force along OF_2 is found in the usual way. The resultant is **OP**, found from

$$OP^2 = 100^2 + 200^2 - 2 \times 100 \times 200 \times \cos O\hat{F}_2P$$

Now if PF_2 is continued to X, $O\hat{F}_2 X = 60°$ (alternate angles) and thus $O\hat{F}_2P = 120°$, so

$$OP^2 = 100^2 + 200^2 - 2 \times 100 \times 200 \cos 120°$$
$$= 100^2 + 200^2 + 2 \times 100 \times 200 \cos 60° \text{ (since } \cos 120° = -\cos 60°)$$
$$= 70\ 000$$

and

$$OP = 264.6$$

The vector difference between the two forces is thus a force of 264.6 N acting along OP. The angle of action, x in the figure, may be determined as in the previous examples by using the sine rule. Thus

$$\frac{200}{\sin x} = \frac{264.6}{\sin 120}$$

and

$$\sin x = \frac{200}{264.6} \times \sin 120$$

$$= 0.6545$$

Hence $x = 40° 53'$ from tables.

USE OF PHASOR DIAGRAMS IN ELECTRICAL CIRCUIT PROBLEMS

Phasor diagrams are used extensively in the solution of problems involving alternating currents and voltages. So far only simple summation of voltages or currents has been demonstrated and little reference has been made to the circuit to which they refer. The remaining examples in this chapter demonstrate the use of phasor diagrams applied to actual series and parallel circuits.

Example 6.12
When a series circuit consisting of a 30 Ω inductive reactance in series with a 40 Ω resistance is connected across an alternating supply voltage of value V_s volts a current of 5 A flows. Draw the phasor diagram showing the component voltages and current and determine the value of V_s and the phase angle between the supply voltage and the circuit current.

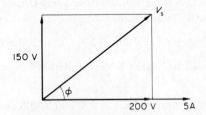

Figure 6.19

The voltage across the inductance is reactance $\times$ current, that is, 30 Ω $\times$ 5 A = 15C and the voltage across the resistor is resistance $\times$ current, that is, 40 Ω $\times$ 5 A = 20(The supply voltage is the phasor sum of these two voltages.

From the diagram shown in figure 6.19

$$V_s^2 = 150^2 + 200^2$$

$$= 62\ 500$$

and

$$V_s = 250\ \text{V}$$

Also $\tan \phi = 150/200$ where ϕ is the phase angle, and $\phi = 36° \ 52'$ from tables.

Example 6.13

A series circuit contains three components: an inductor of reactance $1200\ \Omega$ at the supply frequency and with resistance $200\ \Omega$; a capacitor of reactance $1300\ \Omega$ at the supply frequency; and a resistor of resistance $300\ \Omega$. A current of 2 A flows when a certain a.c. supply voltage is applied. Draw the complete phasor diagram and determine the value of the supply voltage and the phase angle of the circuit.

$$\text{Total inductive reactance} = 1200\ \Omega$$

$$\text{Voltage across inductive reactance} = 1200 \times 2$$
$$= 2400\ \text{V}$$
$$\text{Total capacitive reactance} = 1300\ \Omega$$

$$\text{Voltage across capacitive reactance} = 1300 \times 2$$
$$= 2600\ \text{V}$$

$$\text{Total resistance} = 300 + 200\ \Omega$$
$$= 500\ \Omega$$

$$\text{Voltage across resistance} = 500 \times 2$$

$$= 1000\ \text{V}$$

The phase relationships of these voltages are shown in figure 6.20 where V_s is the (resultant) supply voltage. From the diagram

$$V_s^2 = 1000^2 + 200^2$$

Hence

$$V_s = 1020\ \text{V}$$

and

$$\tan \phi = \frac{200}{1000} = \frac{1}{5}$$

Hence

$$\text{phase angle } \phi = 11° \ 19' \text{ from tables.}$$

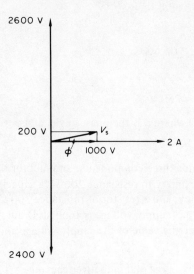

Figure 6.20

Example 6.16

A circuit consisting of a 100 Ω resistor connected in parallel with a capacitor of reactance 75 Ω at the supply frequency is connected to a 300 V alternating current supply. Draw the phasor diagram of the circuit and determine the current in each branch, the supply current and the phase angle between supply current and supply voltage.

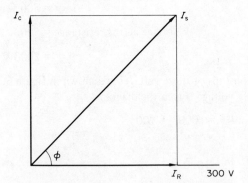

Figure 6.21

The phasor diagram is shown in figure 6.21 in which I_R depicts the current in the resistive branch, I_c depicts the current in the capacitive branch and I_s depicts the supply current. Now

$$I_R = \frac{\text{supply voltage}}{\text{resistance}}$$

$$= \frac{300}{100}$$

$$= 3 \text{ A}$$

and

$$I_c = \frac{\text{supply voltage}}{\text{reactance}}$$

$$= \frac{300}{75}$$

$$= 4 \text{ A}$$

So that

$$I_s^2 = I_R^2 + I_c^2$$

$$= 3^2 + 4^2$$

and

$$I_s = 5 \text{ A}$$

Also

$$\tan \phi = \frac{I_c}{I_R}$$

$$= \frac{4}{3}$$

so that

$$\phi = 53° \ 8' \text{ from tables}$$

The supply current is thus 5 A leading the supply voltage by $53° \ 8'$.

Exercise 6.1

Unless otherwise stated take $g = 9.81 \text{ m/s}^2$ where required.

(1) The string of a bow is drawn back with an arrow such that the string makes an angle of $45°$ on each side of the arrow. If the tension in each part of the string is 200 N calculate the resultant force acting on the arrow when it is released.

(2) An aeroplane travelling at 500 km/h in a direction N $45°$ W is subject to a wind blowing N $10°$ W at 40 km/h. Calculate the resultant speed and direction of the aeroplane.

(3) A body on a smooth surface is subject to a force of 150 N acting along the surface. If the mass of the body is 10 kg determine the magnitude and direction of the resultant force acting on the body.

(4) A resistor of value 150 Ω is connected in series with a capacitor of reactance 200 Ω at the supply frequency, and a current of 2 A flows on connection to an a.c. supply. Determine the supply voltage and the phase angle between supply voltage and current.

(5) Determine the r.m.s. value and phase angle of the resultant of three alternating voltages having instantaneous values given by the following equations

$$v_1 = 100 \sin \omega t$$
$$v_2 = 150 \sin [\omega t - (\pi/3)]$$
$$v_3 = 250 \sin [\omega t + (\pi/2)]$$

(6) A **parallel** circuit has three branches. The instantaneous values of the currents in the branches are given by

$$i_1 = 5 \sin [\omega t + (\pi/4)]$$
$$i_2 = 8 \sin \omega t$$
$$i_3 = 10 \sin [\omega t - (\pi/3)]$$

Determine the equation giving the instantaneous value of the total current.

(7) An inductive coil of resistance 200 Ω is connected in series with a capacitor across a 250 V a.c. supply. If the supply current is 0.2 A and the capacitive reactance at the supply frequency is 200 Ω, calculate the reactance of the inductor at the supply frequency.

(8) Determine the resultant of three voltages of peak value $V_1 = 100$ V, $V_2 = 150$ V and $V_3 = 25$ V, if V_1 and V_2 are in antiphase and V_3 lags their resultant by 40°.

(9) Determine the phase angle between the resultant of the voltages represented by the equations $v_1 = 10 \sin \omega t$, $v_2 = 15 \sin [\omega t - (\pi/3)]$ and $v_3 = 20 \sin [\omega t + (\pi/3)]$ where v_1, v_2 and v_3 are instantaneous values, and the resultant of the currents represented by the equations $i_1 = 2 \sin \omega t$, $i_2 = 4 \sin [\omega t - (\pi/4)]$ and $i_3 = 3 \sin [\omega t + (\pi/4)]$ where i_1, i_2 and i_3 are instantaneous values.

(10) A coil of resistance 20 Ω is placed across a 250 V a.c. supply and a current of 2 A flows. Determine (a) the reactance of the coil at the supply frequency, (b) the phase angle between supply voltage and current.

(11) Calculate the supply current when a parallel circuit consisting of a 50 Ω resistor and a capacitor of reactance 100 Ω is connected across a 100 V a.c. supply.

(12) A circuit is made up of three resistors of resistance 100, 200 and 300 Ω and three capacitors of reactance 50, 100 and 150 Ω (at the supply frequency) connected in parallel. Determine the supply voltage if a supply current of 1 A flows when the circuit is connected across an a.c. supply.

(13) Determine the resultant horizontal and vertical components of the following forces (all in the first quadrant): 100 N acting horizontally; 200 N acting at 40° to the horizontal; 250 N acting at 30° to the vertical; 300 N acting vertically. Hence find the resultant single force.

7 Geometry

In *First-year Technician Mathematics* various properties of triangles, quadrangles and other multi-sided figures were studied and the knowledge of these was applied to the solution of problems. In this chapter some of the more useful theorems concerning circles will be stated and then applied to the solution of further problems. Proofs of these theorems will not be given.

CIRCLES: VOCABULARY OF TERMS

The terms normally used in connection with circles are radius, diameter, sector, segment, arc, chord and tangent. Radius and diameter have already been defined in *First-year Technician Mathematics* while the terms sector and segment have been used without definition in previous chapters.

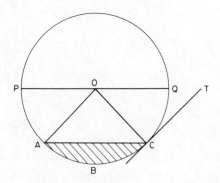

Figure 7.1

Figure 7.1 illustrates the following definitions. In the figure the area OABC is a *sector* of the circle and its boundaries are the *radius* OA, the radius OC and the *arc* ABC. The shaded area ABCA is a *segment* and its boundaries are the *chord* AC and the arc ABC. The segment shown, being less than half the circle area, is called a *minor* segment, the remainder of the area of the circle (the unshaded portion) being called a *major* segment. Any line drawn outside the circle such that it touches the circumference (the circle boundary) only once is called a *tangent*. TC is a tangent to the circle, the point of *tangency* being C, and TC being at right angles to radius OC. OP and OQ are also radii (note the plural) and POQ is the circle *diameter*, which is equal to twice the radius. The diameter may be considered as the circle chord which passes through the circle centre O.

THEOREMS

If two tangents are drawn to a circle from a common point they are of equal length and the angles that the tangents make with the chord joining the points of tangency are equal in value. The line joining the common point to the circle centre bisects the same chord, and also bisects the angle between the tangents.

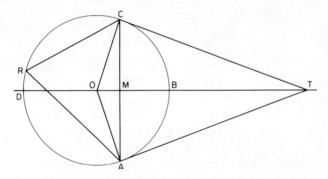

Figure 7.2

Thus in figure 7.2, which shows two tangents TC and TA from common point T, touching a circle at points of tangency C and A, we have

$$TC = TA, \ M\hat{C}T = M\hat{A}T, \ CM = MA \ \text{and} \ C\hat{T}M = M\hat{T}A$$

It can also be shown that the diameter at right angles to a chord bisects it. Thus in figure 7.2, diameter DB is at right angles to CA and the mid-point of CA is M.

If two chords intersect, the products of the parts of each chord produced by the intersection are equal. Thus in figure 7.2 taking DB and AC as chords (DB is actually a diameter which is one particular chord, but the theorem is true for all chords) then DM $\times$ MB = CM $\times$ MA.

If lines are drawn from a common point to the ends of a chord the angle between the lines is said to be *subtended* by the chord. Thus, in the figure AÔC is the angle subtended by chord AC at circle centre O. Similarly AR̂C is the angle subtended by AC at R where R is any point on the circle circumference. It can be shown that all angles subtended by any particular chord at the circumference are equal and that the angle subtended at the circumference by a chord is equal to one-half the angle subtended by the same chord at the centre. Thus, in figure 7.2

$$A\hat{R}C = \tfrac{1}{2} \times A\hat{O}C$$

The angle subtended at the circumference by a diameter is always a right angle.

If two circles touch each other the line joining their centres will always pass through the point of contact as shown in figure 7.3a, but if they intersect as in figure 7.3b this line will bisect the common chord at right angles. In the figure M is the mid point of AC and O_1, O_2 are circle centres.

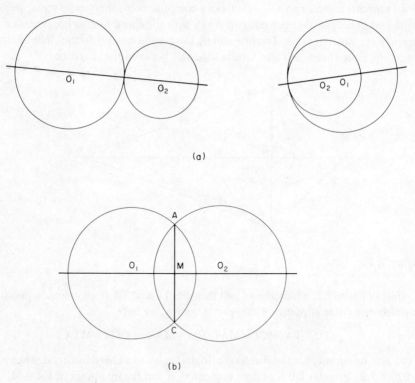

(a)

(b)

Figure 7.3

The theorems stated above should be learnt, since they are most useful in the solution of problems involving circles; they are used in the following examples.

Example 7.1

The cross-section of a machine shaft is in the form of a major arc of a circle of
radius 5 cm cut off by a chord of length 6 cm. Determine the cross-sectional area
of the shaft.

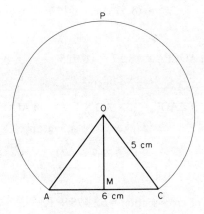

Figure 7.4

The section is shown in figure 7.4. $\triangle AOC$ is an isosceles triangle and M the mid-
point of AC. To find the required area we need to determine the area of the
sector bounded by radii OA, OC and *minor* arc AC (not shown); then subtract
this area from the whole circle area to give area AOCPA and then add to the
answer so far obtained the area of $\triangle AOC$. As shown in *First-year Technician
Mathematics*

$$\text{area of a circle} = \pi \times \text{radius}^2$$

As shown in chapter 4 of this book

$$\text{area of a sector} = \frac{\theta}{2\pi} \times \pi r^2$$

$$= \frac{1}{2} r^2 \theta$$

where θ is the angle between the radii ($A\hat{O}C$ in figure 7.4) in radians and r is the
radius. To find θ, that is, $A\hat{O}C$, first find $M\hat{O}C$ as follows

$$\sin MOC = \frac{MC}{OC}$$

$$= \frac{3}{5} \quad (MC = AC/2)$$

Thus MÔC = 36° 53′ = 0.6437 rad. Hence AÔC = 2 × 0.6437 = 1.2874 rad. Therefor

$$\text{area of sector} = \frac{1}{2} \times 5^2 \times 1.2874$$

$$= 16.0925 \text{ cm}^2$$

$$\text{area of circle} = \pi \times 5^2$$

$$= 78.57 \text{ cm}$$

so that

$$\text{area AOCPA} = 78.57 - 16.925$$

$$= 62.478 \text{ cm}^2$$

$$\text{Area of } \triangle AOC = \frac{1}{2} \times AO \times CO \times \sin AOC \text{ (see chapter 6)}$$

$$= 0.5 \times 25 \times \sin 73° 46′$$

$$= 12.0 \text{ cm}^2$$

therefore

$$\text{Area of cross-section} = 62.478 + 12$$

$$= 74.48 \text{ cm}^2$$

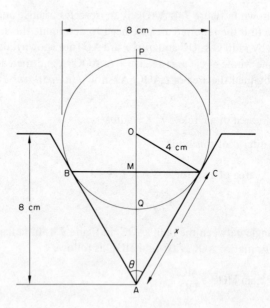

Figure 7.5

Example 7.2

A cylindrical peg rests in a V-cut on a surface plate such that its horizontal diameter is in line with the surface. The radius of cross-section of the peg is 4 cm and the V-cut is 8 cm deep. Calculate (a) the length along the slant surface between the point of contact and the apex of the V-cut;   (b) the angle of the V-cut;   (c) the horizontal distance between slant surfaces of the cut at the points of contact.

The diagram is shown in figure 7.5. Length x is the required length, θ the required angle and BC the required horizontal distance.

(a) To find x consider $\triangle OAC$. AC is tangential to the circle and thus $O\hat{C}A$ is a right angle. Thus

$$x^2 = OA^2 - OC^2 \quad \text{(Pythagoras)}$$
$$= 8^2 - 4^2 \quad \text{(OC is a radius)}$$

and

$$x = \sqrt{48}$$
$$= 6.928 \text{ cm}$$

(b) Angle θ is bisected by the line drawn from the point A to circle centre O since AB, AC are tangents, therefore

$$\sin(\theta/2) = \frac{OC}{OA}$$
$$= \frac{4}{8} = 0.5$$

and

$$\theta/2 = 30° \text{ so that } \theta = 60°$$

(c) To find BC, the theorem regarding intersecting chords may be used for chords BC and PQ (PQ actually being a diameter). Since BM = MC

$$PM \times MQ = BM^2$$

Now

$$MQ = MA - QA$$
$$= MA - 4$$

(since QA = 8 − OQ = 8 − 4) and since

$$MA = x \cos(\theta/2)$$
$$= 6.928 \times \cos 30$$
$$= 6.00$$

thus

$$MQ = 6 - 4$$
$$= 2 \text{ cm}$$

Now

$$PM = PQ - MQ$$
$$= 8 - 2$$
$$= 6 \text{ cm}$$

so that

$$BM^2 = 6 \times 2$$
$$= 12$$

and

$$BM = 3.464$$

thus

$$BC = 2\,BM = 6.928 \text{ cm}$$

Alternatively in $\triangle OMC$ angle $O\hat{M}C$ is a right angle, therefore $CM = OC \cos O\hat{C}M$. But in $\triangle MOC$

$$O\hat{C}M = 90 - M\hat{O}C$$

and since $M\hat{O}C = A\hat{O}C$

$$M\hat{O}C = 90 - (\theta/2)$$
$$= 60$$

thus

$$O\hat{C}M = 90 - 60 = 30°$$

therefore

$$CM = 4 \cos 30$$
$$= 3.464$$

and

$$BC = 2\,CM$$
$$= 6.928 \text{ cm} \quad \text{as before}$$

This alternative method is possibly more straightforward.

Example 7.3

The roller method of determining the radius of a curved surface consists of placing that surface on a horizontal surface and moving in two rollers of known diameter until the curved surface is supported. The arrangement is shown in figure 7.6. In such a method the roller diameter was 10 mm and the separation between roller outer edges, as shown, was found to be 96.4 mm. Calculate the radius of the curved surface.

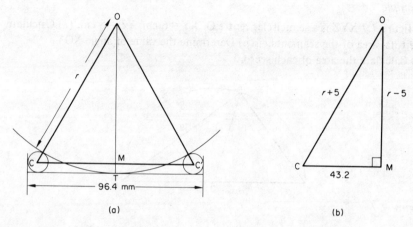

Figure 7.6

Let the curved surface radius be r. The line joining the centres of touching circles passes through the point of contact, thus line OC in the diagram (from curved surface centre O to roller centre C) is equal in length to the radius of the surface r added to the roller radius, that is, $r + 5$ mm. The horizontal surface is tangential to the curved surface and thus OT drawn between centre O and the point of tangency is equal to r in length and is at right angles to the horizontal. If a line joining roller centres is drawn parallel to the horizontal and cutting OT at M, then M is the mid point of line CC′ (congruent triangles OCM, OC′M two sides and a right angle). The length of CM by inspection is $\frac{1}{2}$ × 96.4 less the radius of the left-hand roller, which is 48.2 − 5 mm. The length of OM, which equals OT − MT equals $r − 5$ mm since OT is the curved surface radius r and MT is equal in length to the roller radius. The triangle OCM which is used to calculate r is redrawn as figure 7.6b for clarity. In $\triangle$OCM

$$OC^2 = OM^2 + CM^2 \quad \text{(Pythagoras)}$$

$$(r + 5)^2 = (r - 5)^2 + 43.2^2$$

then

$$r^2 + 10r + 25 = r^2 - 10r + 25 + 43.2^2$$

Collecting terms

$$20r = 43.2^2$$

$$r = \frac{43.2 \times 43.2}{20}$$

$$= 93.31 \text{ mm}$$

Example 7.4

In figure 7.7 XYZ is a semicircle, centre O. XY = 6 cm, YZ = 8 cm. (a) Calculate the total area of the segments.   (b) Determine the value of angle XÔY. (c) Calculate the area of each sector.

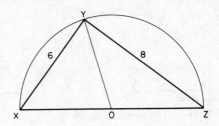

Figure 7.7

(a) The two segments are bounded one by arc XY and chord XY, and the other by arc YZ and chord YZ. The combined area of the segments is equal to the area of the semicircle less the area of triangle XYZ. To find the semicircle area we need to know its radius and for this we use the fact that the angle subtended at the circumference of a circle by the diameter is always a right angle, that is, in this case XŶZ (subtended by diameter XZ) equals 90°. XZ, the diameter, is the hypotenuse of right-angled triangle XYZ, so that

$$XZ^2 = 6^2 + 8^2$$

$$= 100$$

and

$$XZ = 10 \text{ cm}$$

thus the radius OX = OY = OZ = 5 cm.

$$\text{Area of semicircle} = \frac{1}{2} \pi \times \text{radius}^2$$

$$\text{Area of triangle} = \frac{1}{2} \times \text{base} \times \text{height}$$

Thus

$$\text{semi-circle area} = \frac{1}{2} \times \pi \times 5^2$$

area of triangle XYZ = $\frac{1}{2}$ × 6 × 8

so that

area of segments = $\frac{1}{2}$ × π × 5^2 − $\frac{1}{2}$ × 6 × 8

= 15.28 cm²

(b) In triangle XOY, XY = 6 and OY = OX = 5. Using the cosine rule

$XY^2 = OX^2 + OY^2 - 20X \times OY \times \cos X\hat{O}Y$

that is

$6^2 = 5^2 + 5^2 - 2 \times 5 \times 5 \times \cos X\hat{O}Y$

Hence

$\cos X\hat{O}Y = \dfrac{50 - 36}{50} = 0.28$ and $X\hat{O}Y = 73°\ 44'$ from tables

(c) The semi-circle is made up of two sectors, one bounded by arc XY and radii OX, OY and the other bounded by arc YZ and radii OY, OZ. To find the area of a sector we use

$$\text{sector area} = \frac{\text{sector angle (rad)}}{2\pi} \times \pi \times \text{radius}^2$$

$$= \frac{1}{2} \times \text{sector angle (rad)} \times \text{radius}^2$$

The angle of one sector is 73° 44′ or 1.2874 radians, and of the other is 180 − 73° 44′, that is, 106° 16′ or 1.8537 radians. Therefore area of one sector is $\frac{1}{2}$ × 1.2874 × 5^2 which equals 16.09 cm² and the area of the other is $\frac{1}{2}$ × 1.8537 × 5^2 which equals 23.17 cm².

Example 7.5

Two circles of diameters 7 cm and 4.5 cm intersect at X and Y (see figure 7.8). The length of the common chord is 1.2 cm. Determine the distance between centres P and Q of the circles, the area of triangle PYQ, and the angle PYQ.

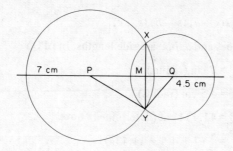

Figure 7.8

The line PQ bisects **XY** at **right angles so that,** letting the point of bisection be M, △PMY is right-angled and △QMY is right-angled. In △PMY

$$PY = 7 \text{ cm (radius)}$$

$$MY = 0.6 \text{ cm (equal to one half of XY)}$$

and since

$$PM^2 = PY^2 - MY^2$$

$$= 7^2 - 0.6^2$$

$$= 48.64$$

and

$$PM = 6.974 \text{ cm}$$

In △QMY

$$QY = 4.5 \text{ cm (radius)}$$

$$MY = 0.6 \text{ cm}$$

and since

$$QM^2 = QY^2 - MY^2$$

$$QM^2 = 4.5^2 - 0.6^2$$

$$= 19.89$$

and

$$QM = 4.46$$

Hence

$$PQ = PM + QM = 11.434 \text{ cm}$$

For triangle PQY all sides are known. The area of a triangle when all sides are known is given by

$$\sqrt{[s(s-a)(s-b)(s-c)]}$$

where $2s$ = sum of sides and a, b, c are side lengths. In △PQY

$$PY = 7 \text{ cm (radius)}$$

$$QY = 4.5 \text{ cm (radius)}$$

$$PQ = 11.434 \text{ cm (determined above)}$$

$$s = \tfrac{1}{2}(7 + 4.5 + 11.434)$$

$$= 11.467 \text{ cm}$$

Thus area of triangle $= \sqrt{[(11.467\,(11.467 - 7)\,(11.467 - 4.5)\,(11.467 - 11.434)]}$

$$= \sqrt{11.78}$$

$$= 3.433 \text{ cm}^2$$

Using the cosine rule on triangle PYQ we have

$$PQ^2 = PY^2 + YQ^2 - 2PY \times YQ \times \cos P\hat{Y}Q$$

so that

$$11.434^2 = 7^2 + 4.5^2 - 2 \times 7 \times 4.5 \cos P\hat{Y}Q$$

and

$$\cos P\hat{Y}Q = \frac{7^2 + 4.5^2 - 11.434^2}{2 \times 7 \times 4.5}$$

$$= -0.9759$$

Hence

$$P\hat{Y}Q = 180 - 12° \ 36' \text{ from tables}$$

(P$\hat{Y}$Q lies in the second quadrant because the cosine is negative and, by examination, it cannot lie in the third.) Therefore

$$P\hat{Y}Q = 167° \ 24'$$

Example 7.6

A sphere rests in a circular hole in a metal plate so that the bottom of the sphere is 4 cm below the plate surface. The hole in the plate has a diameter of 20 cm. Find the sphere radius.

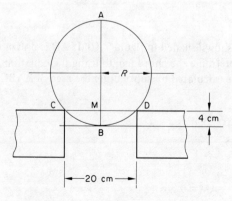

Figure 7.9

The diagram is shown in figure 7.9. AB is a diameter bisecting chord CD at M. Now

$$AM \times MB = CM \times MD \text{ (Theorem)}$$

and since $AM = 2R - MB$, where R is the sphere radius, and $MB = 4$ cm

$$CM = MD = \frac{1}{2} \times 20 \text{ cm}$$

then

$$(2R - 4) \times 4 = 10 \times 10$$

$$2R - 4 = \frac{100}{4} = 25$$

$$R = \frac{29}{2} = 14.5 \text{ cm}$$

Example 7.7

A segment of a circle of radius 9 cm has an arc which subtends an angle of $30°$ at the circle centre. Determine (a) the chord length, (b) the segment area.

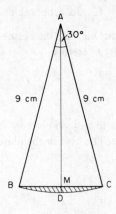

Figure 7.10

The segment is shown shaded in figure 7.10. If AD is drawn to bisect BC at M, then we may determine the chord length using the equation $BM \times MC = AM \times MD$. The area may be calculated by subtracting the area of $\triangle ABC$ from the area of the sector ABDC.

(a) $M\hat{A}C = \frac{1}{2} \times 30°$

$$= 15°$$

Thus

$$AM = 9 \cos 15$$

$$= 8.69 \text{ cm}$$

and

$$MD = 9 - 8.69$$
$$= 0.31$$

so that

$$BM \times MC = BM^2 = 8.69 \times 0.31$$

and

$$BM = \sqrt{(8.69 \times 0.31)}$$
$$= 1.64 \text{ cm}$$

Thus

$$\text{chord length} = 2BM = 3.28 \text{ cm}$$

(b) Area of sector $= \frac{1}{2} \times B\hat{A}C \times AC^2$ (where BAC is in radians)

$$= \frac{1}{2} \times \frac{\pi}{6} \times 9^2$$
$$= 21.2 \text{ cm}^2$$

Area of triangle $ABC = \frac{1}{2} \times AB \times AC \times \sin 30$

$$= 20.25 \text{ cm}^2$$

therefore

$$\text{Area of segment} = 21.2 - 20.25$$
$$= 0.95 \text{ cm}^2$$

Example 7.8

A line OA is drawn from the centre O of a circle of radius 15 cm to a point A situated 15 cm from O. A tangent from A to the circle touches the circle at B. Find the length of AB and the angle $B\hat{A}O$.

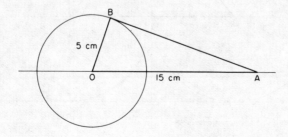

Figure 7.11

In figure 7.11

$$AB^2 = OA^2 - OB^2 \text{ (Pythagoras since O}\hat{B}A = 90°)$$
$$= 15^2 - 5^2$$

Hence

$$AB = 14.14 \text{ cm}$$

Also

$$\sin BAO = \frac{OB}{OA}$$
$$= \frac{5}{15}$$

Thus

$$B\hat{A}O = 19° \ 28' \text{ from tables}$$

Example 7.9

Two bars of equal circular cross-section are laid side by side on a horizontal surface. A third bar of circular cross-section, radius 2.5 cm, is laid on top of the two as shown in figure 7.12. If the centre of the cross-section of the third bar is 10 cm above the horizontal surface calculate the radius of cross-section of the lower bars.

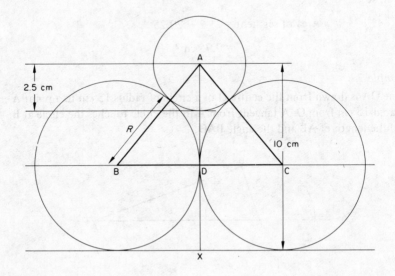

Figure 7.12

In figure 7.12 A, B and C are circle centres. Let R be the radius of the cross-section of the lower bars. The height of A above ground is 10 cm and is equal to AD + DX where, by inspection, DX = R the required radius. The problem therefore reduces to finding AD then solving for R in AD + R = 10. In $\triangle ABC$

$$AB = R + 2.5 \text{ (sum of radii)}$$
$$BD = R$$

And in $\triangle ADB$, $A\hat{D}B = 90°$, so

$$AD^2 = AB^2 - BD^2$$
$$= (R + 2.5)^2 - R^2$$
$$AD = \sqrt{[(R + 2.5)^2 - R^2]}$$
$$AD = 10 - R \text{ (since AD + } R = 10)$$

therefore

$$\sqrt{[(R + 2.5)^2 - R^2]} = 10 - R$$

so that

$$(R + 2.5)^2 - R^2 = (10 - R)^2 \text{ (squaring both sides)}$$

and

$$R^2 + 5R + 6.25 - R^2 = 100 - 20R + R^2$$

that is

$$R^2 - 25R + 93.75 = 0$$

which is a quadratic equation in R. Applying the formula

$$R = \frac{25 \pm \sqrt{(25^2 - 4 \times 93.75)}}{2}$$
$$= \frac{25 \pm \sqrt{250}}{2}$$
$$= \frac{25 \pm 15.81}{2}$$
$$= 20.405 \text{ or } 4.595$$

Clearly the first answer is not possible since if DX (equal to the radius) were 20.405 cm, AX would not be 10 cm. Thus, radius = 4.595 cm.

Example 7.10

Two cylindrical bars of cross-sectional diameters 4 cm and 2.5 cm lie side by side on a horizontal surface as shown in figure 7.13. Calculate the length of the distance x shown.

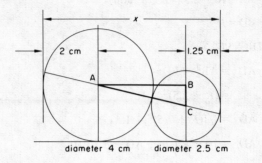

Figure 7.13

In figure 7.13 perpendiculars are drawn through circle centres A and C and AB is drawn horizontally to cut the perpendicular through C at B. Now

$$AC = 2 + 1.25 \text{ cm} \quad \text{(sum of the radii)}$$

$$AB = x - 2 - 1.25 \quad \text{(as shown in figure 7.13)}$$

and BC = 2 − 1.25 since A is 2 cm vertically above the surface and C is 1.25 cm vertically above the surface. Therefore in △ABC

$$AC = 3.25 \text{ cm}$$

$$AB = x - 3.25 \text{ cm}$$

$$BC = 0.75 \text{ cm}$$

and since $\hat{ABC} = 90°$ by construction, in △ABC

$$AC^2 = AB^2 + BC^2 \quad \text{(Pythagoras)}$$

$$3.25^2 = (x - 3.25)^2 + 0.75^2$$

or

$$(x - 3.25)^2 = 3.25^2 - 0.75^2$$

then

$$x^2 - 6.5x + 10.5625 = 10$$

$$x^2 - 6.5x + 0.5625 = 0$$

By the formula

$$x = \frac{6.5 \pm \sqrt{(6.5^2 - 4 \times 0.5625)}}{2}$$

$$= \frac{6.5 \pm \sqrt{40}}{2}$$

$$= \frac{6.5 \pm 6.325}{2}$$

$$= 6.412 \text{ or } 0.087$$

The second answer is not possible because x clearly exceeds the diameter of circle centre A. Thus x = 6.41 cm.

Exercise 7.1

(1) Determine the area of the major segment of a circle of radius 8 cm when the cord length is 10 cm.

(2) A groove of maximum depth 7.5 mm is cut into a flat surface. The cross-section of the groove is a segment of a circle of radius 5 cm. Calculate the groove width.

(3) Two cylindrical bars are laid side by side on a horizontal surface. If the horizontal distance between the outer edges of the bars when viewed in cross-section is 15 cm and the radius of one bar is 3 cm, calculate the radius of the other bar.

(4) Two gear wheels of diameter 25 cm and 10 cm are fixed such that the distance between centres is 30 cm. Calculate the length of drive chain required to connect the wheels, the chain being wrapped around part of each wheel circumference and being taut where it touches neither circumference.

(5) The roller method of determining the radius of a curved surface (see figure 7.6) using rollers of 12 mm diameter is used for a curved surface of radius 25 mm. Calculate the distance between roller centres.

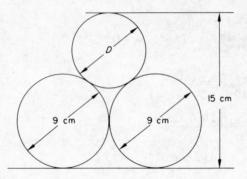

Figure 7.14

(6) Calculate diameter D of the upper circle shown in figure 7.14.

(7) Two tangents are drawn from a point X situated 15 cm from the centre Y of a circle radius 3 cm. Determine (a) the length of each tangent,   (b) the length of the chord joining the points of tangency,   (c) the angle contained by the tangents.

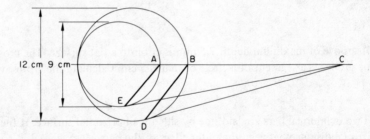

Figure 7.15

(8) In figure 7.15 CD and CE are tangents to circles of diameter 12 cm and 9 cm respectively situated as shown. Length BC is 20 cm. Calculate (a) EA,   (b) BD, (c) EÂC,   (d) DB̂C,   (e) CD,   (f) CE.

(9) A fan blade is a sector of a circle of radius 14.3 cm such that the sector area is one eighth of the area of the complete circle. Calculate the angle subtended at the circle centre by the sector arc, the area of the sector and the length of the chord joining the radii ends.

(10) Three holes A, B and C drilled in a metal plate have their centres on the circumference of a circle of diameter 9.7 cm. If hole A is at a distance of 5.2 cm from hole B and 9.7 cm from hole C find the distance between holes B and C.

(11) A circle of radius 8 cm has two chords XY and VW that intersect at point Z such that XZ = 3.4 cm, ZY = 1.5 cm and VZ = 7 cm. Find the length ZW and the area of segment VYWZ.

(12) A sphere of diameter 237 mm rests in a circular hole in a metal plate so that the bottom of the sphere is 120 mm below the plate surface. Find the diameter of the hole.

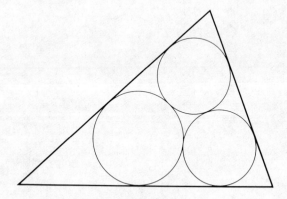

Figure 7.16

(13) Three cylindrical bars of diameters 15 mm, 15 mm and 20 mm are arranged as shown in figure 7.16. Determine the length of the sides of the enclosing triangle.

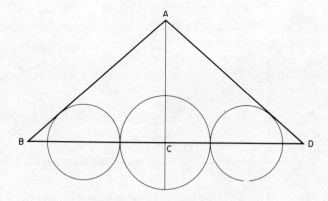

Figure 7.17

(14) The diameter of the smaller circles in figure 7.17 is 13.5 cm and of the larger circle is 16.4 cm. Find the area of the triangle shown if the length of AC is 25 cm.

(15) Two concentric circles are drawn and tangents to the upper part of each circle constructed from a point 29 cm from the centre. If the diameter of the smaller circle is 4 cm and the length of the tangent to the larger circle is 24 cm, find the diameter of the larger circle and the angle between the tangents.

Test Paper 1

(1) Simplify

(a) $\left(\dfrac{16}{81}\right)^{-3/4}$ (b) $\left(\dfrac{16384}{78125}\right)^{-1/7}$

(2) If $x/y = \sqrt{[(m - n)/(m + n)]}$ show that

$$\frac{m}{n} = \frac{y^2 + x^2}{y^2 - x^2}$$

Find m/n if $y = 3.7x$.

(3) The values of two variable quantities u and v are observed to be as follows

u	4.6	6.4	9.4	13.6	19
v	1	2	3	4	5

Show that the equation relating u and v is of the form $u = kv^2 + c$ (where k and c are constants) and determine the values of k and c.

(4) Find the values of x which satisfy the equation $3x^2 = 4x + 13$ correct to two decimal places.

(5) In a series-connected electrical circuit the voltage across one component leads the current by $30°$ and has a value 150 V, the voltage across the other component lags the current by $90°$ and has a value 175 V. Determine by calculation the resultant voltage and its phase relationship with respect to the current.

(6) If

$$y = \frac{x}{2}\left[m + (x-1)r\right]$$

express r in terms of x, m and y.

(7) Find the values of θ between $0°$ and $360°$ which satisfy the equation $\cos\theta = \sqrt{2}\sin\theta$.

(8) If $\sin x = 4/5$ and $\cos y = 5/13$ find the value of $\cot x + \tan y$.

(9) In triangle XYZ, XY = 3.5 cm, XZ = 4.2 cm and $X\hat{Y}Z = 97°$. Calculate the remaining angles, the length YZ and the triangle area.

(10) Simplify

$$\frac{(x^2 - y^2)(x + y)^2}{(x\,y)(x - y)}$$

(11) Solve for x the equation $(x + 4)\,/2 = x + 4$.

(12) An inverted cone of base diameter 10 cm and slant height 15 cm is filled with liquid to a depth of 5 cm from the cone apex. Find the volume of liquid.

(13) Find the value of

$$\frac{(1 + \cot^2 x)\sin 2x}{\operatorname{cosec}^2 x} \times \sqrt{0.0041}$$

when $x = 192° \, 41'$.

(14) (a) Convert the binary number 110.1011 to denary.
      (b) Use binary addition and subtraction to determine the value of the following

$$1011 + 101101 + 110 - 1011 - 1011$$

Express the answer in denary form.

(15) A voltage rises linearly from zero to 10 V in 0.2 seconds, remains at 10 V for the next 0.2 seconds, continues to rise linearly to 20 V during the next 0.2 seconds and then falls linearly to zero in 0.2 seconds. Use the mid-ordinate rule to find the average value of the voltage over the whole period of 0.8 seconds.

Test Paper 2

(1) (a) The collector current I_c of a transistor is related to the collector voltage V_c by an equation of the form $I_c = kVc + c$ where k and c are constants. When $V_c = 5$ V, $I_c = 4.1$ mA and when $V_c = 10$ V, $I_c = 4.23$ mA. Find the values of k and c.

(b) The values of two variables x and y are observed and a graph plotting y^3 against x results in a straight line of gradient 0.43 and passing through the origin. Write down the equation connecting the variables.

(2) Plot the graphs of $y = \sin 2\theta$ and $y = \tan \theta$ for values of θ between 0 and 180°. Hence solve the equation $\sin 2\theta = \tan \theta$ for values of θ in the given range.

(3) (a) Make f the subject of the formula $Z = \sqrt{[R^2 + (2\pi fL)^2]}$.

(b) Simplify $10x^{-4} y^{-3} \div 5x^2y^3$ leaving only positive indices in the answer.

(4) Solve for x the equation $y = (x - 3)(x + 1) - 14$ expressing the answer to two places of decimals.

(5) Factorise $2x^2 + 5x - 12$ and $2x^2 - 5x + 3$ and use the results to simplify

$$\frac{1}{2x^2 + 5x - 12} - \frac{1}{2x^2 - 5x + 3}$$

expressing the answer as a single fraction.

(6) Solve the equations

(a) $0.784^{x + 2} = 0.891$

(b) $\log_{10} (x - 4) = 0.3142$

(7) XYZ is a semi-circle, diameter XZ and Y lying on the circumference. The circle centre is at 0. If XY = 5 cm and YZ = 12 cm calculate the circle diameter and the area of each circle segment.

(8) The depth of a river is measured at three-metre intervals to give the following results

Distance from bank (m)	0	3	6	9	12	15	18	21	24
Depth (m)		0	1.2	2.4	4	6.1	4.8	3.5	2	0

The river is 24 m wide. Use the mid-ordinate rule to determine (a) the average depth of the river   (b) the cross-sectional area.

(9) A cylindrical container holds 4.7 litres of liquid when full. If the container height is 3.2 times the diameter determine the diameter of the cylinder.

(10) (a) Express in binary notation (i) 315   (ii) 0.1374 to seven places.

(b) Convert to denary the binary number 101101.1101.

(11) In an a.c. circuit four voltages V_1, V_2, V_3 and V_4 have the following phase relationships and magnitudes

V_1 is 150 V peak

V_2 is 32.4 V r.m.s. lagging V_1 by 35°

V_3 is 48.9 V peak leading V_2 by 90°

V_4 is 112.4 V r.m.s. lagging V_3 by 10°

Find the resultant voltage by resolving into components in phase and in quadrature with V_1.

(12) Using tables evaluate (a) cos 281° 14′   (b) cosec 179° 14′   (c) cot $5\pi/12$
(d) sec (1.43 rad).

(13) The angles of a triangle are in the ratio 1: 2: 3; the longest side is 13.5 cm; calculate the length of the shortest side.

(14) The angle of elevation of the top A of a vertical radio mast AB, height 120 metres, from a point C due south of B is 37°. A point D is located N 41° E of C at a distance of 300 metres. If B, C and D are on the same horizontal level calculate the angle of elevation of A from D and determine the distance between D and the foot of the mast B.

(15) Solve the simultaneous equations

$$3.1x = 5 - y$$
$$x + 2.7y = 7$$

Test Paper 3

(1) Prove the identity $\sin \theta + \cos \theta = (1 + 2 \sin \theta \cos \theta)^{1/2}$

(2) (a) Transpose the formula $L = \mu_0 N^2 A / l$ to make N the subject.

(b) Find the value of N if $L = 10^{-3}$, $\mu_0 = 4\pi \times 10^{-7}$, $l = 0.35$ and $A = 30 \times 10^{-6}$.

(3) (a) If $\log_{10} 3 = x$ and $\log_{10} 7 = y$ express $\log_{10} 63 / \log_{10} 147$ in terms of x and y without using tables.

(b) Solve the equation $\log_2 (x + 1)^2 = 5$ for x.

(4) The values of two variable quantities m and n are observed to be as follows

m	0.0019	0.023	0.041	0.063	0.081
n	0.4472	0.6325	0.7746	0.8944	0.1

The equation relating m and n is believed to be of the form $m = kn^2 + c$. Test this by graphical means and find values of k and c if the relationship is correct.

(5) Solve the following equations for values of θ between 0 and $360°$

(a) $\tan^2 \theta = 0.94$ (b) $\sin 2\theta = 0.987^{1/2}$

(6) A radar station A is situated 320 km due north of another radar station B. An unidentified flying object is located at a bearing S $23°$ W from A and N $32°$ W from B. Find the distance of the object from A and from B.

(7) Solve the following quadratic equations for x to *two* decimal places

(a) $x - 5 = \dfrac{8}{x + 1}$

(b) $\dfrac{x - 7}{4} = \dfrac{1}{x + 2}$

(8) An electric current changes linearly from zero to 5 A in 0.1 second, remains at that value for 0.15 second then reduces linearly to zero in 0.1 second. Determine the r.m.s. value of this current using Simpson's rule.

(9) Three voltages v_1, v_2 and v_3 have instantaneous values given by the equations

$$v_1 = 215 \sin 3142t$$
$$v_2 = 170 \sin [3142t - (\pi/6)]$$
$$v_3 = 115 \sin [3142t + (\pi/3)]$$

Determine the peak values, r.m.s. values, frequency and periodic time of these voltages. Draw the phasor diagram and determine the resultant voltage by calculation.

(10) (a) $\dfrac{12.3 \times 23.6 \times 896 \times 25.1 \times 0.141}{5.6 \times 231 \times 0.1456}$

The above expression when calculated on a slide rule gives the figures 489. By reducing each figure to standard form and *without calculation* find the position of the decimal point.

(b) Express denary 836 in binary. (c) Express 1101.1101 (binary) in denary.

(11) Two circles, each of diameter 8.3 cm, intersect so that the length of the common chord is 2.7 cm. Find the distance of the common chord from each circle centre.

(12) (a) Simplify $(2^{1/3})^4 \times (2^4)^{1/3} \div (8^2) \times (16^{1/3})$

(b) If $4^{-2/3} = x$ and $2^{-1/3} = y$ show that $(xy)^{-1/2} = 2^{5/6}$.

(13) Plot the graphs of $y = 4x + 1$ and $y = x^2 - 3x + 2$ between $x = -2$ and $x = +7$. Hence find the roots of the equation $x^2 - 7x + 1 = 0$.

(14) A sphere rests in a circular hole in a flat surface such that the bottom of the sphere lies 0.5 cm below the surface. Find the sphere diameter and its volume.

(15) Use tables to evaluate (a) $\cos 342° \, 18'$; (b) $\tan 756° \, 17'$; (c) $\sin 1341° \, 15'$; (d) $\operatorname{cosec} 972° \, 14'$.

Test Paper 4

(1) If $x = \sin^2\theta - 1$ and $y = \sec^2\theta - 1$ show that

  (a) $x(y + 1) = -1$ and (b) $x + 1 = -xy$

(2) Use tables to evaluate

$$\text{(a) } \sqrt{(0.00143)} \qquad \text{(b) } \sqrt{\left(\frac{1}{0.0.346}\right)}$$

(3) Draw a graph of $y = x^2 - 2x + 7$ from $x = -1$ to $x = 3$ and find the area under the curve using Simpson's rule.

(4) Two variables m and n are related by the equation $m = an^{1/2} + b$ where a and b are constants. Observed values of m and n are as follows

m	3.8	4.4	5	5.6	6.2
n	1	4	9	16	25

Plot a suitable graph and determine the values of a and b.

(5) Sove the following equations for x to two decimal places

  (a) $3x^2 + 4x - 5 = 0$ (b) $0.4x^2 - 3.1x - 2.9 = 0$

(6) Make ω the subject of the formula $Z = \sqrt{[R^2 + (\omega L - \omega C)^2]}$.

(7) Find the values of θ between $0°$ and $360°$ which satisfy the equation $\sin^3\theta = \sqrt{0.987}$.

(8) The instantaneous values i_1, i_2 and i_3 of three alternating currents in a parallel connected circuit are given by the following

$$i_1 = 5 \sin \theta$$
$$i_2 = 6 \sin [\theta + (\pi/6)]$$
$$i_3 = 7 \sin [\theta - (\pi/5)]$$

Draw the phasor diagram and determine the components of the resultant currents which lie in phase and $90°$ out of phase with i_2.

(9) Solve the equations

(a) $0.519^{x^2} = 0.037$ (b) $\ln (x + 1)^{1/2} = 3.571$

(10) A ship steaming due east at 48 km/h is sighted twice by an observer. The first time the ship is at a bearing N $31°$ $25'$ W; and the second time, 1 hour and 45 minutes later, at a bearing N $25°$ $12'$ E. Find the shortest distance between ship and observer during this period.

(11) Two similar cylindrical bars having equal diameters are placed side by side on a horizontal surface, the circular cross-section being in a vertical plane. A third bar having a circular cross-section of diameter 7 cm is placed on top of the first two bars so that the centre of its cross-section lies 5.4 cm vertically above the surface. Calculate the diameter of the lower bars.

(12) A current i changes its magnitude in accordance with the equation $i = 2 + 0.3t$ over the time period $t = 0$ seconds and $t = 2$ seconds, then falls linearly to zero in 1.6 seconds. Find the r.m.s. value of the current over the total period.

(13) A metal cone of base diameter 4.7 cm and slant height 3.1 cm is melted down and re-cast without waste in the form of a cylinder of height 2.7 cm. Find the diameter of the cylinder.

(14) Evaluate the following using tables

(a) $\sec 722°$ (b) $\cot 1135°$ (c) $(\sec 273°)^{1/2}$

(15) Convert (a) 31.742 denary to binary (to six places);   (b) 1101110.1101 binary to denary.

Test Paper 5

(1) Use tables to evaluate

    (a) $\sec 114° \ 13'$ (b) $\operatorname{cosec} 923° \ 10'$ (c) $\cot 512° \ 50'$

(2) Prove the identities

    (a) $\sin \theta = (1 + 2 \sin \theta \cos \theta)^{1/2} - \cos \theta$

    (b) $\dfrac{\sin \theta}{1 + \cos \theta} = \dfrac{1 - \cos \theta}{\sin \theta}$

(3) The values of two variables x and y are observed to be as follows

y	5.16	5.32	5.48	5.64	5.80
x	0.0256	0.4096	2.0736	6.5536	16

Show that x and y are related by an equation of the form $y = k_1 x^{1/4} + k_2$ where k_1, k_2 are constants; determine the values of k_1 and k_2.

(4) Find the area of a minor segment of a circle radius 4 cm, chord length 2.5 cm, using Simpson's rule.

(5) Solve for x to two places of decimals

    (a) $\dfrac{4}{x - 3} = \dfrac{x + 4}{5}$ (b) $\dfrac{x - 5}{3} = \dfrac{7}{x}$

(6) Make R_1 the subject of the formula

$$\frac{1}{R_T} = \frac{R_1 + R_2}{R_1 R_2} + \frac{1}{R_3}$$

(7) Find the values of θ between $0°$ and $360°$ which satisfy the equation

$$\frac{1 - \cos^2\theta}{\sin\theta} = \frac{1}{\sqrt{14.74}}$$

(8) In a series circuit voltage V_1 has a maximum value of 85 V and leads voltage V_2 by $42°$. V_2 has an r.m.s. value of 27.5 V. Voltage V_3, which has a peak to peak value of 112.5 V lags V_1 by $71°$. Draw the phasor diagram and calculate the r.m.s. value of the resultant voltage and the phase angle it has with V_2.

(9) Solve the following equations for x

(a) $1 - 0.071^{x-1} = 0.742$ (b) $x^{0.071} = 0.859$

(10) A vertical mast is supported by two ropes tied to the mast top and anchored at ground-level. The mast and ropes are all contained in the same vertical plane. If the distance between rope-anchorage points is 8 m and the angles the ropes make with the ground are $32°$ and $41°$ find the mast height and the length of each rope.

(11) Tangents AB and AC are drawn from a point A to a circle centre O. If the length of each tangent is equal to twice the radius, find the angle between tangents, $B\hat{A}C$, and the angle subtended at the circle centre by chord BC.

(12) A voltage V changes from zero in accordance with the equation $V = 0.4t$ (where t is time in seconds) over a period of 0.1 second, remains constant for 0.2 second then falls linearly to zero in 0.1 s. Find the average value of the voltage over this time period using the mid-ordinate rule.

(13) Find the value of

$$\frac{(\sin\theta + \cos\theta)^2 - 2\sin\theta\cos\theta}{\sin^2\theta}$$

when $\theta = 153° \ 14'$.

(14) A spherical container of diameter 5 m is half full of a chemical that is then transferred to a container consisting of an inverted cone of depth 3.4 m. If the second container is completely filled by the liquid, find its diameter at the liquid surface.

(15) Convert (a) 31.72 denary to binary (to six places);   (b) 11011.110 binary to denary.

Answers

CHAPTER 1

Exercise 1.1

(1) (a) 11011   (b) 111000101   (c) 11011100010   (d) 100100101100101
(2) (a) 59   (b) 107   (c) 3003   (3) (a) 1110100 (116)   (b) 111110 (62)
(c) 100101011 (299)   (d) 1001 (9)   (4) (a) 1001111   (b) 100000011
(c) 10001   (5) (a) 0.110   (b) 101.100   (c) 1101.001   (6) (a) 0.90625
(b) 14.75   (c) 22.84375   (7) (a) 1101.001   (b) 110000.0001   (8) (a) 10010.11
(b) 1000.1   (9) 1.10011   (10) 10001.1101

Exercise 1.2

(1) 0.0010284   (2) 0.2264552   (3) 0.0068   (4) 0.3110   (5) 1.2008
(6) 6.1387   (7) 0.4964   (8) 0.2885   (9) 2.412

Exercise 1.3

(1) (a) 0.8506   (b) 3.1532   (c) 5.4558   (d) 7.7584   (2) (a) 0.8506
(b) 3.1532   (c) 5.4558   (d) 7.7584   (3) (a) 3.99 A   (b) 3.84 s
(4) (a) 4.76 V   (b) 10.94 V 69.3 s   (5) 0.9377   (6) (a) $x = 11.87$
(b) $y = 4.497$ (7) (a) 2.893   (b) 4.1388   (c) 3.2865   (8) (a) 3.4426
(b) $\overline{2}.6213$ (9) (a) 2.1   (b) 96.54   (10) (a) 7.042   (b) 0.7042
(c) 0.07042   (d) 0.007042   (11) (a) 23.71   (b) 0.0319
(12) 525.2

CHAPTER 2

Exercise 2.1

(1) $y = z^{2/5}$ (2) 32 768   (3) $b = a^{-2/15}$ (4) $x^{1/2}y^{1/5}z^{1/3}$ (5) 0.1323
(6) $\frac{2}{9}x^3y^2z^2$ (7) 12.317   (8) $m^{2/30}n^{-13/10}p^{-5/8}$ (9) 2.08
(10) $r^{7/6}/st^{5/4}$

Exercise 2.2

(1) $C = 1/4\pi^2 f_r^2 L$ (2) $G_L = (h_r h_f - h_o h_i + h_o R_{IN})/(h_i - R_{IN})$
(3) $R_L' = Z_{od}/(1 - Z_{od}g_{fs})$ (4) $h_{fb} = (h_{ob} - h_{oc})/h_{oc}$ (5) $\Phi = \dfrac{\pi a T}{pZI_a}$

(6) (a) $\tau = \dfrac{1}{t}\left[\dfrac{\log 1/(1 - V_R/E)}{\log e}\right]$ (b) $\tau = \dfrac{1}{t}\ln\left(\dfrac{1}{1 - V_R/E}\right)$

(7) $C = \dfrac{L}{X_L^2 + R_L^2}$ (8) $R_L = \sqrt{\left(\dfrac{L}{C} - 4\pi^2 f_r^2\right)}$ (9) $I_m = \sqrt{\left(\dfrac{I_c^2 - I_c^2\cos^2\theta_0}{\cos^2\theta_0}\right)}$

(10) $\lambda_2 = \frac{1}{2}\left\{-\lambda_1 \pm \sqrt{[\lambda_1^2 + 4\lambda_1^2/(m_p\lambda_1 - 1)]}\right\}$ (solution of quadratic in λ_2)

Exercise 2.3

(1) 0.808 A, 0.085 A   (2) 0.543 A or 0.109 A (determined by battery connection)
(3) (a) $I_1 = 1.7933, I_2 = 0.655$ (b) $I_1 = 14.386, I_2 = 5.263$ (4) $x = 3, y = 3/5$
(5) $V = 120$ km/h   (6) $c = 15$ (7) $a = 2.4, b = 6$ (8) $P_1 = 11.18, P_2 = 3.471$
(9) $V_1 = 5.25, V_2 = 1.125$ (10) $k = 1.298, c = 1.8226$

Exercise 2.4

(1) (a) $-4, -1$ (b) $6, -1$ (c) $-5, 4$ (d) $\frac{1}{2}, -\frac{1}{4}$ (e) $-3, \frac{1}{2}$ (f) $-\frac{4}{9}$
(2) (a) $+1, -3$ (b) 8.873, 1.127   (c) $-0.3882, -0.6118$ (d) 3.5
(e) 0.6449. 0.1551   (f) $-1.275, -3.7247$ (3) (a) $0.405, -7.405$ (b) 1,
-0.4 (c) 5.7837, 0.2162   (d) $0.809, -0.309$ (e) $1.1546, -2.9372$
(f) $1.9608, -0.9104$ (4) $R_1 = 12.825, R_2 = 6.075$ or $R_1 = 6.075, R_2 = 12.825$
(5) 0.566 A   (b) 30 m   (7) 103.21 V   (8) 5.464   (9) $-0.45, -15.54$
(10) 3.47 s   (11) 10 V   (12) 1 V   (13) 3.271 cm   (14) 238.75 Hz
(15) 4.7287

CHAPTER 3

Exercise 3.1

(1) $x = 4$ and 3 (2) $x = 1$ and $-\frac{1}{2}$ (3) $x = -1$ and 2 (4) (a) $x = 7$ and -1
(b) $x = 4$ and $\frac{3}{2}$ (5) $x^2 + 6x + 5 = 0$ (6) $x = -\frac{1}{2}$ and $-\frac{1}{4}$ (7) $x = 5.464$
and -1.464 (8) $x = 7.3$ and -9.3 (9) $x^2 - 5x - 84 = 0$ (10) $x = 3$ and -3

Exercise 3.2

(1) $V = 7$ (2) $V = 10, R_2 = 6.1$ (3) $k = 1, C_S = 5 \, \mu\text{F}$ (4) $k = 125, R_S = 3 \, \Omega$
(5) $k = 3$ (6) Graph is y against $\sqrt{x}, a = 2, b = 5$ (7) $q = 0.7p + 1.1$
(8) $a = 0.045, b = 4.8$ (9) $a = 50, b = 30$ (10) $k = 31.42$ (11) $k = 2500$,
$m = 100$ (12) $a = 0.8, b = 8$ (13) $k = 8 \times 10^{-5}$ (14) $k = 0.3, n = 2$
(15) $k = 95.6, n = 0.667$ (17) $c = 527.3, n = 1.42$
(18) $a = 1.79 \times 10^{-3}, b = \frac{5}{2}$ (19) $y = 0.23x^{1.33}$ (20) $k_1 = 0.09, k_2 = 0.0055$

CHAPTER 4

Exercise 4.1

(1) 44   (2) 24.67   (3) 2587.2 m³/min   (4) 0.6146 mm, 2.3895 mm³
(5) 35.67 V, 39.285 V   (6) 7.07 V   (7) 7   (8) 472.5 J   (9) Average 15 V;
r.m.s., 21.21 V; form factor and crest factor 1.414   (10) Average 1.8137 A;
r.m.s. 2.176 A   (11) 10.6 V   (12) 9.192 A   (13) Mean 12 V; r.m.s. 16.97 V
(14) 10.863 V

Exercise 4.2

(1) 1.584 × 10⁶ cm³/minute   (2) 2.464 m³   (3) 1.164 m   (4) 34.96 cm³;
23.96 cm³   (5) 3101.78 mm²   (6) 36.132 cm³   (7) 35.68 cm   (8) 19.37 cm,
10 cm   (9) 6.448 cm   (10) 68.30 cm²; 29.46 cm³   (11) 0.9232 cm; 88.14%
(12) 2.466 cm   (13) 7.07 minutes; 95.75%   (14) 471.43 cm²

CHAPTER 5

Exercise 5.1

(1) 0.8870   (2) -0.4838 (3) -1.5951 (4) -1.1274 (5) -2.0669
(6) -0.6269 (7) -0.2090 (8) -0.5601 (9) -5.1446 (10) 0.9848
(11) -0.9998 (12) -1.3269 (13) 0.4067   (14) -0.6157 (15) 3.1210
(16) -1.0006 (17) 1.2522   (18) 0.5317   (19) 0.9998   (20) -0.1219

Exercise 5.2

(1) 3.4 A; 2.4 A; 100 Hz; 10 ms   (2) $v = 25 \sin 133.33 \, \pi t$; $- 17.11$ V
(3) 36.53 W   (4) (a) 25; 17.678; 15.909; 50 Hz   (b) 4; 2.828; 2.545; 100 Hz
(c) 115; 81.317; 73.182; 25 Hz   (d) 25; 17.678; 15.909; 318.2 Hz
(e) 7.9; 5.586; 5.027; 500 Hz   (f) 27.5; 19.445; 17.5; 159.1 Hz   (g) 4; 2.828;
2.545; 100 Hz   (h) 0.1; 0.0707; 0.0636; 50 Hz   (i) 2.5; 1.768; 1.591; 100 Hz
(j) 25; 17.678; 15.909; 100 Hz   (5) $11° \, 44'$; 98.03 V   (6) $i_1 = 10 \sin 200\pi t$,
$i_2 = 5 \sin (200\pi t + \pi/12)$; 0.4851 ms, 0.9017 ms   (7) $v = 145.46 \sin (100\pi t +$
$0.173)$ angle in radians   (8) $i = 7 \sin (50\pi t + 0.38)$ angle in radians
(9) $v = 26.87 \sin 133.33\pi t$; $i = 7 \sin (133.33\pi t + 0.2667\pi)$

Exercise 5.3

(12) 2   (13) 0.07111   (14) 0.03627

Exercise 5.4

(1) $270°, 210°, 330°$ (2) $194° \, 29', 345° \, 31'$ (3) $36° \, 52', 323° \, 8'$
(4) $53° \, 8', 126° \, 52', 233° \, 8', 306° \, 52'$ (5) $231° \, 40', 308° \, 20'$ (6) $0°$,
$104° \, 29', 255° \, 31'$ (7) $110° \, 44', 249° \, 16'$ (8) $38° \, 11', 141° \, 49'$
(9) $64° \, 16', 295° \, 44', 219° \, 53', 140° \, 7'$ (10) $10° \, 42', 79° \, 17'$

Exercise 5.5

(1) (a) $c = 8.9$, $\hat{A} = 40° \, 8'$, $\hat{B} = 104° \, 52'$ (b) $a = 3.944$, $\hat{B} = 28° \, 26'$, $\hat{C} = 123° \, 34'$
(c) $c = 5.586$, $\hat{B} = 31° \, 53'$, $\hat{A} = 138° \, 7'$ (d) $b = 6.56$, $\hat{A} = 60° \, 55'$, $\hat{C} = 84° \, 25'$
(e) $c = 16.09$, $\hat{A} = 47° \, 37'$, $\hat{C} = 52° \, 23'$ (f) $b = 52.98$, $\hat{C} = 14° \, 6'$, $\hat{B} = 148° \, 54'$
(g) $c = 36.24$, $\hat{B} = 31° \, 11'$, $\hat{C} = 133° \, 49'$ (h) $b = 23.48$, $\hat{C} = 38° \, 8'$, $\hat{B} = 89° \, 52'$
(i) $c = 29.22$, $\hat{A} = 13°$, $\hat{C} = 154°$ (2) Height 5 m; distance 19.57 m
(3) 41.46 km, 36.63 km   (4) 9.3674 cm^2 (5) 357.59 km, N $81° \, 23'$ W
(6) 43.72 m   (7) 29.369 cm   (9) $90° \, 54'$ (10) $26° \, 23'$ (11) 2.073 cm^2
(12) 6.479 cm^2

CHAPTER 6

Exercise 6.1

(1) 282.84 N   (2) 533.26 km/h at N $42° \, 32'$ W   (3) 179.2 N at $33° \, 11'$ down
from horizontal   (4) 500 V lagging current by $53° \, 8'$ (5) 150.17 V r.m.s.,
$34° \, 28'$ lead   (6) $i = 17.31 \sin (\omega t - 0.3)$ angle in radians   (7) 1433.9 Ω

(8) 70.99 V lagging 13° 5'   (9) voltage leads by 14° 45'   (10) (a) 123.38 Ω
(b) 80° 48'   (11) 2.236 A   (12) 269.19 V   (13) 747.75, 59° 37'

CHAPTER 7

Exercise 7.1

(1) 189.14 cm²   (2) 52.68 mm   (3) 3.97 cm   (4) 117.06 cm   (5) 32.5 mm
(6) 3.675 cm   (7) (a) 14.697 cm   (b) 11.51 cm   (c) 23° 4'   (8) (a) 5.82 cm
(b) 7.43 cm   (c) 130° 16'   (d) 128° 17'   (e) 25.3 cm   (f) 27.13 cm
(9) 45°, 80.33 cm², 20.22 cm   (10) 8.188 cm   (11) 0.7286 cm, 32.28 cm²
(12) 236.98 mm   (13) 59.86 cm, 59.86 cm, 35.33 cm   (14) 609.63 cm²
(15) 32.56 cm, 30° 12'

TEST PAPER 1

(1) (a) 27/8   (b) 5/4   (2) 1.1576   (3) $k = 0.6, c = 4$ (4) 2.852, $- 1.519$
(5) 82° 35' lag   (6) $(2y - mx)/[x(x - 1)]$ (7) 35° 16', 215° 16'   (8) 63/20
(9) 55° 47', 27° 13', 2.36 cm, 6.078 cm²   (10) $(x^2/y) + 3x + 3y + (y^2/x)$
(11) $- 4, - 3$ (12) 16.369 cm³   (13) 0.0292   (14) (a) 6.6925   (b) 40
(15) 10 V

TEST PAPER 2

(1) (a) $k = 0.13, c = 3.97$ (b) $y^3 = 0.43x$ (2) 45°, 135°, 225°, 315°
(3) (a) $f = \sqrt{[(z^2 - R^2)/4\pi^2 L^2]}$ (b) $2/x^6 y^6$ (4) 3.47 or $- 5.47$
(5) $- 3/(2x - 3)(x + 4)(x - 1)$ (6) (i) $- 1.526$ (ii) 6.062   (7) 13 cm,
1.686 cm², 34.709 cm²   (8) (a) 3 m   (b) 72 m²   (9) 1.231 cm
(10) (a) (i) 100111011   (ii) 0.0010001   (b) 45.8125   (11) 306.58 V peak
leading 56'   (12) (a) 0.1948   (b) 74.6269   (c) 0.2679   (d) 7.0972
(13) 4.5 cm   (14) 29° 59', 207.97 m   (15) $x = 0.882, y = 2.266$

TEST PAPER 3

(2) (a) $N = \sqrt{(lL/\mu_0 A)}$ (b) 3046   (3) (a) $(2x + y)/(2y + x)$ (b) 4.657
(4) $k = 0.1, c = 0$ (5) (a) 44° 7', 135° 53', 224° 7', 315° 53'   (b) 41° 43'
48° 17'   (6) 204.5 km from A, 150.81 km from B   (7) (a) 6.12, $- 2.12$
(b) 7.42, $- 2.42$ (8) 4.226 A   (9) Peak values 215 V, 170 V, 115 V; r.m.s.
values 152.03 V, 120.21 V, 81.32 V; frequency 500 Hz, periodic time 2 ms; Resultant is
419.97 V peak leading 2° 70'   (10) (a) 4890   (b) 1101000100   (c) 13.8125
(11) 3.924 cm   (12) (a) $2^{-14/3}$ (13) 6.85 or 0.1459   (14) 18.5 cm,
3316.6 cm³   (15) (a) 0.9527   (b) 0.7342   (c) $- 0.9883$ (d) $- 1.05$

TEST PAPER 4

(2) (a) 0.0378   (b) 5.3760   (3) 29.33   (4) $a = 0.6$, $b = 3.2$ (5) (a) 0.79,
-2.12 (b) 8.59, -0.84 (6) $\omega = (Z^2 - R^2)^{1/2}/(L - C)$ (7) 86° 12', 93° 48'
(8) 13.177 A (in phase), 8.894 A (90° out of phase)   (9) (a) ± 5.0274   (b) 0.8897,
-2.8897 (10) 77.68 km   (11) 1.007 cm   (12) 2.119 V   (13) 2.348 cm
(14) (a) 1.0002   (b) 0.7002   (c) 1.3551   (15) (a) 11111.101111
(b) 110.8125

TEST PAPER 5

(1) (a) -2.4378 (b) -2.5419 (c) -1.9486 (3) $k_1 = 0.4$, $k_2 = 5$
(4) 0.338 cm^2 (5) (a) 5.179, -6.179 (b) 7.72, -2.72
(6) $R_1 = R_2 R_3 R_T/(R_2 R_3 - R_2 R_T - R_T R_3)$ (7) 15° 6', 164° 54', 195° 6',
346° 54'   (8) 108.98 V leading 11° 4'   (9) (a) 1.5122   (b) 0.0575   (10) Rope
lengths; 5.489 m, 4.433 m; mast height: 2.908 m   (11) $\hat{BAC} = 53°$ 8', angle
subtended $(\hat{BOC}) = 126°$ 52'   (12) 0.03 V   (13) 1.25147   (14) 24.25 cm
(15) (a) 11111.101110   (b) 27.75